[H.A.S.C. No. 115–53]

HEARING

ON

NATIONAL DEFENSE AUTHORIZATION ACT FOR FISCAL YEAR 2018

AND

OVERSIGHT OF PREVIOUSLY AUTHORIZED PROGRAMS

BEFORE THE

COMMITTEE ON ARMED SERVICES HOUSE OF REPRESENTATIVES

ONE HUNDRED FIFTEENTH CONGRESS

FIRST SESSION

___

SUBCOMMITTEE ON TACTICAL AIR AND LAND FORCES HEARING

ON

COMBAT AVIATION MODERNIZATION PROGRAMS AND THE FISCAL YEAR 2018 BUDGET REQUEST

___

HEARING HELD
JUNE 7, 2017

U.S. GOVERNMENT PUBLISHING OFFICE

26–738 WASHINGTON : 2018

___

For sale by the Superintendent of Documents, U.S. Government Publishing Office
Internet: bookstore.gpo.gov   Phone: toll free (866) 512–1800; DC area (202) 512–1800
Fax: (202) 512–2104   Mail: Stop IDCC, Washington, DC 20402–0001

SUBCOMMITTEE ON TACTICAL AIR AND LAND FORCES

MICHAEL R. TURNER, Ohio, *Chairman*

FRANK A. LoBIONDO, New Jersey	NIKI TSONGAS, Massachusetts
PAUL COOK, California, *Vice Chair*	JAMES R. LANGEVIN, Rhode Island
SAM GRAVES, Missouri	JIM COOPER, Tennessee
MARTHA McSALLY, Arizona	MARC A. VEASEY, Texas
STEPHEN KNIGHT, California	RUBEN GALLEGO, Arizona
TRENT KELLY, Mississippi	JACKY ROSEN, Nevada
MATT GAETZ, Florida	SALUD O. CARBAJAL, California
DON BACON, Nebraska	ANTHONY G. BROWN, Maryland
JIM BANKS, Indiana	TOM O'HALLERAN, Arizona
WALTER B. JONES, North Carolina	THOMAS R. SUOZZI, New York
ROB BISHOP, Utah	(Vacancy)
ROBERT J. WITTMAN, Virginia	
MO BROOKS, Alabama	

JOHN SULLIVAN, *Professional Staff Member*
DOUG BUSH, *Professional Staff Member*
NEVE SCHADLER, *Clerk*

CONTENTS

Page

STATEMENTS PRESENTED BY MEMBERS OF CONGRESS

Tsongas, Hon. Niki, a Representative from Massachusetts, Ranking Member, Subcommittee on Tactical Air and Land Forces 3

Turner, Hon. Michael R., a Representative from Ohio, Chairman, Subcommittee on Tactical Air and Land Forces 1

WITNESSES

Bunch, Lt Gen Arnold W., USAF, Military Deputy, Office of the Assistant Secretary of the Air Force for Acquisition; and Lt Gen Jerry D. Harris, USAF, Deputy Chief of Staff for Plans, Programs, and Requirements 6

Grosklags, VADM Paul A., USN, Commander, Naval Air Systems Command, U.S. Navy; LtGen Jon M. Davis, USMC, Deputy Commandant of the Marine Corps for Aviation, U.S. Marine Corps; and RADM DeWolfe H. "Chip" Miller III, USN, Director of the Air Warfare Division, U.S. Navy 4

APPENDIX

PREPARED STATEMENTS:
Bunch, Lt Gen Arnold W., joint with Lt Gen Jerry D. Harris 87
Grosklags, VADM Paul A., joint with LtGen Jon M. Davis and RADM DeWolfe H. "Chip" Miller III 28
Turner, Hon. Michael R. 25

DOCUMENTS SUBMITTED FOR THE RECORD:
[There were no Documents submitted.]

WITNESS RESPONSES TO QUESTIONS ASKED DURING THE HEARING:
Mr. Langevin 115
Ms. McSally 116

QUESTIONS SUBMITTED BY MEMBERS POST HEARING:
Mr. Bacon 124
Mr. Gaetz 123
Ms. Tsongas 123
Mr. Turner 121

(III)

COMBAT AVIATION MODERNIZATION PROGRAMS AND THE FISCAL YEAR 2018 BUDGET REQUEST

House of Representatives,
Committee on Armed Services,
Subcommittee on Tactical Air and Land Forces,
Washington, DC, Wednesday, June 7, 2017.

The subcommittee met, pursuant to call, at 3:33 p.m., in room 2118, Rayburn House Office Building, Hon. Michael R. Turner (chairman of the subcommittee) presiding.

OPENING STATEMENT OF HON. MICHAEL R. TURNER, A REPRESENTATIVE FROM OHIO, CHAIRMAN, SUBCOMMITTEE ON TACTICAL AIR AND LAND FORCES

Mr. TURNER. The hearing will come to order.

The subcommittee meets today to review the Navy, Marine Corps, and Air Force aviation investment and modernization budget request for fiscal year [FY] 2018.

I would like to welcome our distinguished panel of witnesses. We have Vice Admiral Paul Grosklags, Commander of the Navy Air Systems Command; Rear Admiral "Chip" Miller, Director of the Navy's Air Warfare Division; Lieutenant General Arnold Bunch, Military Deputy in the Office of the Assistant Secretary of the Air Force for Acquisition. We have Lieutenant General Jerry Harris, Air Force Deputy Chief of Staff for Plans, Programs, and Requirements. And we have Lieutenant General Jon Davis, Deputy Commandant of the Marine Corps for Aviation.

I understand this will be General Davis' final appearance before the committee because he will be retiring next month. General Davis, we want to thank you for your 37 years of distinguished service to the Marine Corps and our Nation. And we wish you the best in your future plans. And I want you to know that I know everyone on this committee has greatly appreciated the assistance that you have provided us because you give us not only just a perspective on what we should be doing, but also your work has been incredibly inspirational. So thank you for your work.

General DAVIS. My honor, sir. Thanks.

Mr. TURNER. I also would like to thank all of you for your service and look forward to your testimony today.

As I have stated at previous hearings, I support the President's commitment to rebuilding the capacity and capability of our military. However, I am concerned that the current budget request of merely $603 billion for the Department of Defense will not achieve that goal in the timeline desired and needed.

For example, at Congress' request, the military services submitted their unfunded requirements list to the congressional defense committees last week. The total was over $30 billion. A significant portion of these requests were related to modernization needs. Of particular note, the Air Force included an additional 14 F–35A aircraft, and the Navy and Marine Corps included 20 additional aircraft comprised of F–18 Super Hornets, F–35Bs, and F–35Cs.

I suspect that all the witnesses today will support the President's budget request. However, members of this subcommittee need to better understand what additional capabilities are required above the President's request and why it is an imperative we work to fully resource these unfunded requirements to accelerate the restoration of full-spectrum readiness.

I continue to support Chairman Thornberry and Chairman McCain, who believe that a $640 billion budget in fiscal year 2018 is required to build the capability needed for today's complex and dangerous world.

The military services' unfunded requirements also validate a higher topline funding level. Today, the subcommittee will review a broad portfolio of tactical aviation modernization programs and associated acquisition strategies. The witnesses have been asked to identify their top five modernization requirements and briefly summarize how the budget request addresses them.

The aviation budgets for the Navy, Marine Corps, and Air Force appear to be placing a higher priority on current readiness, and rightly so. Earlier this year, we heard the Vice Chief of Naval Operations report that over 60 percent—I am going to say that over again—60 percent of F–18s are out of service due to backlogs in depot repair. The budget fully funds depot capacity.

This budget also robustly funds preferred aviation munitions, such as the Joint Direct Attack Munition, Small Diameter Bombs, and Hellfire missiles—a much needed increase, especially for the ongoing combat operations in the U.S. Central Command's [CENTCOM's] areas of responsibilities. However, the Navy continues to absorb the significant risk in its management of the strike fighter inventory. The Navy is challenged to replace legacy F–18C, D, and AV–8B aircraft that have reached the end of their life service before they can be replaced by new F–35s or new F–18 Super Hornet aircraft. Each year, the Navy flies about 180,000 flying hours in its F–18 fleet, which equates to the entire fleet expending 24 to 36 aircraft worth of service life per year.

The Navy's fiscal year 2018 budget request includes a procurement of 38 F–18 Super Hornets and F–35s, so the Navy is only slightly above its annual expenditure of fighter aircraft life. There are two less Navy F–35Cs in the budget request than were projected last year. And the Future Years Defense Program for fiscal year 2018 reduces the planned F–35C procurements by seven aircraft.

The Air Force currently has a fleet of 55 combat-coded fighter squadrons, significantly smaller than the Desert Storm force of 134 fighter squadrons. We have heard testimony from senior Air Force leaders that with the current demand for rotational fighter presence, 55 combat fighter squadrons do not allow sufficient time to train pilots, maintain aircraft, which contributes to degraded full spectrum readiness. The 55 combat squadron level meets the minimum requirement set forth in the current defense planning guidance, but the Air Force considers it to be high risk in many challenging scenarios.

The Air Force has stated that in order to meet current steady-state demands and maintain readiness to meet surge requirements, the Air Force needs to grow at least 60 combat fighter squadrons, invest in munitions, modernize existing platforms, and increase participation in advanced training programs.

The fiscal year 2018 budget request includes 45 F–35As for the Air Force, and that number is 2 more than planned last year for this budget request. However, I would also note that 2 years ago, the Air Force planned to procure 60 F–35As in the fiscal year 2018.

In testimony before this committee last July, Air Force General "Hawk" Carlisle, the former commander of Air Combat Command, testified that to address the Air Force's capability capacity shortfalls the desired production rate is 60 F–35s per year, not the 46 that this request includes.

The committee is also pleased to see the Air Force reverse its decision to retire the U–2 in 2019 and provide funding to maintain both the U–2 and Global Hawk platforms to meet high-altitude airborne intelligence, surveillance, and reconnaissance [ISR] demands for our combat commanders. However, all of the services represented today need to take a hard look at their investment strategies and airborne ISR capacity, because a significant portion of combat commanders' requirements are still unmet in many of the intelligence disciplines.

As the committee continues its deliberations on the fiscal year 2018 budget request, we will look for opportunities to further address the services' most promising and pressing modernization requirements.

As you know, that is an incredibly depressing list of gaps that is going to be the focus of our hearing today. It is unusual, actually, for this long of an opening statement, but I think each of those elements were important to set the stage for the testimony that we are going to be receiving today, because we are not just focused on what you are going to accomplish and what you have been requesting, but the gap of what needs to be there and how it has been falling short and what we are losing as a result of that continuing gap.

And with that, I would like to turn to my good friend and colleague from Massachusetts, Niki Tsongas.

[The prepared statement of Mr. Turner can be found in the Appendix on page 25.]

STATEMENT OF HON. NIKI TSONGAS, A REPRESENTATIVE FROM MASSACHUSETTS, RANKING MEMBER, SUBCOMMITTEE ON TACTICAL AIR AND LAND FORCES

Ms. TSONGAS. Thank you, Mr. Chairman. And good afternoon to our witnesses. Thank you for your service. And especially General Davis, we wish you well in your retirement. It has been a pleasure to work with you.

The budget request for Air Force, Navy, and Marine Corps programs in our jurisdiction reflects a more stable funding path for major aviation programs and seeks to address modernization upgrades needed by all three services. This committee has rightfully focused on areas where improvements are needed in the acquisition process, but there are a number of successes that don't get much attention, but promise to deliver greatly improved capabilities in some key areas in the next 2 years for the services. Our witnesses today and the thousands of people that work in acquisition for the services should be commended for that.

That being said, there are areas of concern with some of the programs being reviewed by the subcommittee today. And it is our responsibility to make sure our service members get the best possible equipment and to see that taxpayer funds are wisely spent.

Today's hearing will have to cover a lot of ground, but I wanted to make two quick points prior to our hearing—to hearing from our witnesses. The first is my continued concern with the rate of physiological events plaguing the Navy's aviation community in the F/A–18 Hornet and T–45 trainer aircraft fleet. While I know the Navy is working hard and looking at lots of options in this area, the rates remain at a very high rate in some cases. And in the case of the T–45C, the Navy hasn't been able to conduct student pilot flights since early April, a more than 2-month pause in critical training for our future aviators.

I look forward to getting an update on the issue and to hearing about how the several efforts underway are addressing the problem. I also have some questions about areas for Air Force and Navy collaboration to solve these issues.

My second concern is with the significant cost growth reflected in the budget request for several important programs for each of the services. In each case, there are significant increases compared to last year's projections for fiscal year 2018. The programs in question include: one, the Air Force's next-generation [next-gen] air dominance program, which shows a significant increase from $21 million in fiscal year 2017 to $294 million in fiscal year 2018. Second, the Navy's Next Generation Jammer program, which shows an $89 million increase in fiscal year 2018, the F–35 program that shows a $500 million increase in fiscal year 2018 to finish development, and the CH–53K program that shows a $107 million increase in fiscal year 2018.

All of these are very important programs, so I would like to fully understand the reasons for the proposed growth as part of the subcommittee's review of the fiscal year 2018 budget.

And with that, I yield back.

Mr. TURNER. Without objection, all witnesses' prepared statements will be included in the hearing record. Our presenters will be Admiral Grosklags and General Bunch. We will begin with the Admiral.

STATEMENT OF VADM PAUL A. GROSKLAGS, USN, COMMANDER, NAVAL AIR SYSTEMS COMMAND; LTGEN JON M. DAVIS, USMC, DEPUTY COMMANDANT OF THE MARINE CORPS FOR AVIATION; AND RADM DeWOLFE H. "CHIP" MILLER III, USN, DIRECTOR OF THE AIR WARFARE DIVISION

Admiral GROSKLAGS. Thank you, sir. Chairman Turner, Ranking Member Tsongas, and distinguished members of the subcommittee. I appreciate the opportunity to appear before you today to talk about naval aviation programs.

Our 2018 President's budget submission is governed by Secretary of Defense's priorities to improve warfighting readiness by addressing pressing programmatic shortfalls that have accrued from 15 years of wartime operational tempo and chronic underfunding of many of our readiness accounts.

The budget request is designed to maintain the operational effectiveness of our current force while also building a bridge to growing the future force starting in 2019. Current readiness of our naval aviation forces is clearly less than it needs to be. The fiscal year 2017 enacted budget provided much needed increases in funding for many of our naval aviation readiness accounts. Our fiscal year 2018 request builds on 2017 with a request for funding of these readiness accounts that both in real terms and as a percent of the requirement is at a level not seen in 8 to 10 years.

Support for these accounts is the most important single lever in returning our fleet to the required levels of readiness. Close behind that is the need to continue, and in some cases accelerate, the procurement of new aircraft. This includes F–35s, as you mentioned, Mr. Chairman, for both Marine Corps and the Navy, and additional F–18 Super Hornets for the Navy. As we continue to struggle with extending the service life and maintaining the readiness of our legacy F–18s, both services are working to accelerate the transition to other aircraft.

In addition to the F–35B and C models, critical priorities for the Marine Corps include initiation and ramp-up of the CH–53K production, completing procurement of the KC–130J, execution of the V–22 common configuration readiness and modernization initiative, and initiation of the MAGTF's [Marine air-ground task force] expeditionary unmanned air system. Each of these priorities is a key contributor to the Marine Corps' capability and capacity to meet both OPLANs [operation plans] and combatant commander requirements.

On the Navy side of the house, in addition to the F–18s and F–35s that I mentioned, pushing forward with the MQ–4 Triton procurement, awarding a development contract for the MQ–25 carrier-based unmanned tanker aircraft, continuing on-track development of the Next Gen Jammer system, and fielding of the long range anti-ship missile on both the B–1 and the F–18 Super Hornet are our priorities.

We will continue to leverage every tool and opportunity available to drive down the cost of each of these programs. This subcommittee has been very supportive of our efforts in the past, and we are again asking for your support for initiatives such as the F–35 block buy and a third V–22 multiyear program supporting the final 7 years of planned Marine, Navy, and Air Force procurements.

Separate from the procurement focus, as mentioned by Ranking Member Tsongas, this subcommittee is very aware of the continued challenges we face in resolving the high rate of physiological episodes that we have seen in our T–45s and our F–18s. It bears repeating that this is naval aviation's number one safety issue, and we continue to approach root cause assessment and near-term mitigation steps from an unconstrained resources perspective.

As we continue to assess potential root causes, in parallel, we are focused on implementation of aircrew alerting and protection systems so that we can resume student training in our T–45s and continue mitigation of risk for our F–18 air crew.

Naval aviation's priorities are directly tied to increasing worldwide security challenges. Our ability to achieve the improved readiness, the increased capacity, and the enhanced capabilities required to deal with these challenges remains constrained by the overall resourcing constraints imposed by the Budget Control Act and the often inefficient use of resources driven by seemingly chronic extended execution under continuing resolutions [CRs].

You have our commitment to making the best possible use of the resources we are given. We ask this subcommittee's continued support in working to eliminate these barriers. We want to thank you again for your continued support of our sailors and marines, and we look forward to addressing your questions.

[The joint prepared statement of Admiral Grosklags, General Davis, and Admiral Miller can be found in the Appendix on page 28.]

Mr. TURNER. General Bunch.

STATEMENT OF LT GEN ARNOLD W. BUNCH, USAF, MILITARY DEPUTY, OFFICE OF THE ASSISTANT SECRETARY OF THE AIR FORCE FOR ACQUISITION; AND LT GEN JERRY D. HARRIS, USAF, DEPUTY CHIEF OF STAFF FOR PLANS, PROGRAMS, AND REQUIREMENTS

General BUNCH. Thank you, Chairman Turner, Ranking Member Tsongas, and other members of the subcommittee, for the opportunity to appear before you today. We appreciate your service and the support this subcommittee provides the United States Air Force, our airmen, and their families.

For the past 70 years, from the evolution of the jet aircraft to the advent of the ICBM [intercontinental ballistic missile], satellite-guided bombs, remotely piloted aircraft, and many other accomplishments, your Air Force has been breaking barriers as a member of the finest joint warfighting team on the planet. For the last 27 of those 70 years, we have been in continuous combat. During this period, we employed air power in ways never envisioned and delivered unparalleled support to the combatant commanders, our sister services, allies, and coalition partners.

While providing this unmatched operational capability, budget realities have taken a toll on our ability to provide for the future joint force. These many years of combat have taught us much, most importantly, that the demand for air power has grown in every mission, in every domain, and in every location. The world has watched your Air Force operate and the world has adapted. Our adversaries have adapted their capabilities to strike at areas we depend on to execute our missions, adapted their defenses to reduce our ability to employ our forces, and adopted many of our tactics and techniques, all of which reduce our ability to employ our forces.

Today, we face a world of ever-improving adversaries, increasing threats, and a persistent war against violent extremism. This changing environment of increasing demands and commitments, along with a limited pool of resources to address issues, and the threat of the Budget Control Act, make our mission of providing unmatched Global Vigilance, Global Reach, and Global Power ever more challenging.

The result of these changes is a marked decrease in our technological advantage. Where I once would have said we had a decided advantage on all fronts, today I can say we retain our lead in some technological areas. However, in other areas, our potential adversaries are nipping at our heels or are shoulder to shoulder with us.

To address the shrinking technology gap, we must continue to invest in science and technology and modernize our forces to ensure our most valued treasure, America's sons and daughters, has a decisive advantage when we send them into harm's way. We do not want a fair fight.

The FY 2018 budget we submitted is the best balance of our readiness and modernization we could achieve within the physical constraints we face. We take this balanced approach seriously, as we must be ready for today while simultaneously preparing for tomorrow's challenges.

The budget request you received continues our emphasis on recovering readiness, filling critical gaps, and improving lethality. The budget invests heavily in airmen, readiness, nuclear deterrence operations, space and cyber capabilities, combat air forces, and infrastructure. It supports the end strength growth we need to start to address combatant commanders' requirements, while also focusing on pilot production, a national crisis for us.

We continue to maintain and modernize the nuclear enterprise, while also prioritizing the resiliency, future capabilities, and modernization of our space domain to operate in increasingly contested domains, environments.

The budget also supports research, development, and fielding of game-changing technologies. As a department, we had to make tough choices in balancing capability, capacity, and readiness, while focusing on modernizing weapon systems and infrastructure. These decisions were not made easily or taken lightly, highlighting that unfulfilled requirements remain.

As you are aware, the budgetary needs of the United States Air Force exceed projected topline funding as demand for Air Force capabilities currently far exceeds our supply. Uncertainty looms over the Department as sequestration and budget—and the Budget Control Act caps return with this year's budget. Budget stability remains vital, and relief of the Budget Control Act limits is necessary for the Air Force to realize its long-term strategy and meet today's and tomorrow's demands. If the law does not allow relief, it could lead to a repeat of the negative consequences of sequestration seen in FY 2013.

We request your engagement and assistance to ensure we do not go down that path again. General Harris and I look forward to answering the committee's questions today. Thank you again for the opportunity, and thank you for your service.

[The joint prepared statement of General Bunch and General Harris can be found in the Appendix on page 87.]

Mr. TURNER. Thank you, gentlemen.

Luckily, everyone in this subcommittee and our committee are opposed to CRs and sequestration, so we all have the same theme and are working in the same direction.

We are going to have votes on the House floor, so we have somewhere between 45 minutes and an hour in order to accomplish everyone having an answer to a question. We are going to limit the time for a question and answer to 3 minutes so we can get to everyone. We did this last time and it seemed to work pretty effectively. We are doing it for the ranking member and the chair also. So I would appreciate if you would assist us in giving both complete, but also short answers. I think within that 3-minute time period, if we get to a point where someone has got to elaborate, you know, certainly we can continue.

But my question goes to General Davis, Admiral Miller, and General Harris, and this is where you get to do a commercial. I appreciate everything you have said about the budget as submitted, but we are very concerned about the unfunded mandates. As we know, the Navy and Marine Corps, over 60 percent of the Department of Navy F–18s cannot fly. The Air Force says it is the oldest and, you know, worst equipped in history that we have had. There is $3.9 billion that is asked for.

General Davis, Admiral Miller, General Harris, if we put that $3.9 billion back, can you tell what we get and what we miss if we don't? General Davis.

General DAVIS. Sir, those are F–35s, both Bs and Cs, or F–22s, C–130s, V–22s, H–1 Zulu attack helicopters, and some OSA, operational support aircraft, airframes. Absolutely positively have to have that new inventory on the line. As we said, you know, the number's in the 60 percent that—today I can fly 91 F–18s airborne. I have got 171 requirement. So if you looked at that 75 percent mission capable rate, I should have 128 that you should be able to fly any given day, so roughly we are 27 short today, this morning. That is three squadrons worth of airplanes.

So the bottom line is it is an imperative for us to recapitalize the force to get the new metal on the line, and that additional money will allow us to do that.

Mr. TURNER. Admiral.

Admiral MILLER. Yes sir, should additional funds above the President's budget be made available, the Navy's FY 2018 unfunded priority list predominantly accelerates the recovery and the readiness and wholeness of the entire fleet, which was the premise of the 2018 budget. What you will notice on that Navy's unfunded priority list is aircraft procurement there at the top of the list, specifically with strike fighters, F–18s and F–35s. So those additional 10 aircraft for the F–18s and 4 aircraft for the F–35s replenish combat-worn aircraft to reduce near-term strike fighter shortfalls and address long-term inventory deficits. As you know, we are fighting inventory management issues. These aircraft absolutely address that.

For the F–35 specifically, it also accelerates our squadron transition plan and gets much needed fifth-generation [fifth-gen] capability onto our aircraft carriers. Lower in the unfunded priority list you will notice some enabler accounts, the things that we talk about that are key to readiness: spares, logistics, and support. We were unable to fully fund to 100 percent those accounts. The unfunded priority list does that for us. Thank you.

Mr. TURNER. General Harris.

General HARRIS. Sir, first on our list would be readiness, and that gets at the airmen that we need to support and make sure that we have the growth for the Air Force based on the task given to us. The priority of those would be in our maintenance, our operations, and also our acquisition forces. The readiness modernization portion of that, it's the additional F–35s to get us to 60 per year, which we think is a minimum to get after the fifth-gen requirements that we have. You will also see some for nuclear deterrent ops to include our NC3 [nuclear command, control, and communications] operations and our ability to command and control our nuclear forces along with space and improving what we are doing on orbit both in our defenses and our SBIRS [Space Based Infrared System] satellite.

So it is a large spectrum, but at the bottom of that list or still on the list is the infrastructure, because for a long time we have been cutting back on our infrastructure improvements and our ability to launch the power projection areas that we need within our Air Force.

Mr. TURNER. Ms. Tsongas.

Ms. TSONGAS. Thank you.

I appreciate in your testimony, Admiral Grosklags, that you mentioned the efforts you are putting into addressing the physiological episodes. But as of today, it has been 2 months since normal training operations using the T–45 trainer aircraft were underway at Kingsville, Meridian, and Pensacola Naval Air Stations. So this is clearly a major problem for the Navy. And the long-term impact will certainly be more pronounced, the longer issues with the trainers persist. So without a reliable flow of new pilots and naval flight officers, the Navy will be in serious trouble from a personnel standpoint in the future.

So in that regard I have two questions about possible ways forward. First, assuming this problem continues for some time, are there non-T–45 aircraft options for getting this training accomplished? If so, what are they? Can Air Force aircraft be used for at least some of the training syllabus? And can the training with the T–45 be cut back and then made up later in an aviator's career? So that's one.

Second, several of the aviators our staff spoke with asked the Navy to take a serious look at changing at least some T–45 aircraft back to a liquid oxygen [LOX] based system for providing breathing air to the crew. And they pointed out that the original British version of the aircraft did use such a system. So is the Navy looking at this option for T–45s? How might this time and cost—what might the time and cost be for doing that? And how can we in Congress help make sure this gets a serious look and testing in the near future? So a lot for a very short time.

Admiral GROSKLAGS. Okay, yes, ma'am. Let me start with the second question first on LOX. We are pursuing LOX as a potential solution to the problem. Today, we have got kind of a two-pronged approach. One is to try and identify root cause and then take corrective actions. The other is looking at alerting and protection measures so we can get the T–45s back in the air for flight training. And we are doing those two things in parallel. One is not waiting for the other.

LOX is, quite honestly, a longer term solution. We are trying to come up with alert and protection methodologies that would get our students back in the aircraft in the near term. We are talking weeks instead of months that we would be looking at trying to put LOX in that aircraft. But we are still pursuing that as an alternative.

Given the constraints, I won't talk about the alert and protection devices that we are looking at putting in the aircraft in the near term unless you want to follow up with that.

In regards to T–45s and doing the training in other aircraft, we have looked and are continuing to look as this continues to stretch whether or not there are things we can push to the next stage of training in the fleet readiness squadrons. We prefer not to do that, obviously, because it costs more in those fleet jets. We are having our own readiness issues with those fleet aircraft, so putting more workload and more burden of ours in those squadrons is—I won't say last resort, but it is not something we want to pursue unless we absolutely have to.

We have considered using other aircraft, potentially Air Force. I don't think we have talked to the Air Force about that. But one of the key things that you get out of T–45 training is taking our new aviators and the student naval aviators to the ship, which we can only do right now in the T–45 or in their next fleet aircraft. One of the challenges is determining what their next fleet aircraft is as they are going through the training syllabus. We would have to make those decisions before they completed their normal——

Mr. TURNER. Admiral, we need you to conclude.

Admiral GROSKLAGS [continuing]. Syllabus training.

So, yes, ma'am, we are looking at all those options, including LOX, but we—again, we are more focused right now at trying to get the T–45s back in the air with alert and protection measures for our student naval aviators.

Ms. TSONGAS. Thank you.

Mr. TURNER. Mr. LoBiondo.

Mr. LOBIONDO. Thank you.

General Bunch, for you, the 177th Fighter Wing in my district has Block 30 F–16s that are in urgent need of AESA [active electronically scanned array] radar upgrades. Luckily, phase 1 and 2 of the AESA joint urgent operational need is now considered fully funded. Additionally, 2-year $243.9 million contract to build 72 radars for the Air Force was just awarded. The contract allows procurement of roughly 450 to 500 radars and the program of record is set at 300. However, there is no additional pre-block F–16 aircraft in the modernization plan for the radars.

I am concerned that if additional funding is not provided for the pre-block F–16, alert facilities like the one I represent at the 177th will only have eight uniquely configured aircraft per squadron. In the proposed F [FY] 2018 budget appears there is a little over $40 million for the F–16 phase 3 research and development. Could you speak to the need for the future procurement funding for the AESA radar so that the Air Force can upgrade the entire fleet, including the remainder of the pre-block F–16s?

General BUNCH. Yes, sir, thank you for that. And you are right, we did award a contract to Northrop Grumman in May. We are starting the program to be on track for what we committed to, which was an IOC [initial operating capability] of third quarter FY 2019, and for the three units tasked for defending the National Capital Region to have that fully completed by the first quarter of FY 2021.

Our procurement funding for additional aircraft starts—and that is to get to 72 aircraft that you mentioned, sir—our procurement funding starts in FY 2020 to address our requirement for 300 more radars. We do have, within the constraints of the contract, the room to grow that based on the numbers that you have said that we have 450 plus that we could put in that. Right now, that is the money we have got laid in for those procurements, so about 300 more.

Mr. LoBIONDO. Okay. Thank you. I yield back.

Mr. TURNER. Ms. Rosen.

Ms. ROSEN. Thank you. I want to thank you, Chairman Turner, Ranking Member Tsongas. I want to thank all the panelists for being here today.

Of course, I represent southern Nevada. We have Nellis Air Force Base, Nevada Test and Training Range, Creech Air Force Base down south. So I am concerned about the procurement of the F–35s and, of course, the length of time that it takes, specifically as it relates to Nellis and our Red Flag training exercises.

You know, all of those exercises, they are critical, those first 10 missions, critical training for our pilots to get out there and be battle ready. So how does this—what are you going to do to maintain a healthy fighter force while you are ramping up with procurement?

General HARRIS. Well, ma'am, we will continue—thank you for that question. Nellis is one of the crown jewels of the U.S. Air Force and the airspace it has provided and all the support that comes from the community. That is our premier Air Force training location for our advanced training, which is perfect for the F–22s that are there and also the F–35s as the fleet grows. So we continue to acquire as many as we can in a year so that we can replace the fleets of older fourth-generation aircraft. Hopefully, we will have fewer to modernize in the future and make sure that we get the training for both fifth-gen primary, but also pushing that training back to our fourth-gen teammates that play there, whether it is in the weapons school effort or in our Red Flag.

Ms. ROSEN. Thank you. I yield back.

Mr. TURNER. Mr. Knight.

Mr. KNIGHT. Thank you, Mr. Chair. And real quickly, General Davis, thank you very much for your candor over the years. Give us an idea what the F–35 is bringing to you in Marine aviation.

General DAVIS. Thanks so much for that question. It is an honor to serve.

Ma'am, also, the Marines are up there at Nellis. We have instructors now integrating with the United States Air Force Fighter Weapons School, both in the test unit up there, and we like going up to the Red Flag as well. And we like our Air Force and Navy brothers and Army brothers coming down to fly with us down there in Yuma at our weapons school.

I will tell you the F–35Bs are now forward deployed and, actually, they are permanently stationed in Japan, out there at the forward edge, at the tip of the spear. The combatant commander asked for the F–35s to fly in Korea a short while ago as a message to those who would try to do our Nation or our allies harm.

Fantastic capability. Red Flag Alaska again up there as well. What we are finding that this airplane is doing for us it really is, I think, changing the way that we fight and changing the way we are going to fight for the foreseeable future. The young aviators that are flying the airplane are finding all kinds of things to do that we didn't even dream of. So it is a—and we are doing with that airplane, we are going to target areas that we never thought possible before. They see things they didn't see before. They are able to process data, and the kill ratios and the kill rates, both the air-to-air and the air-to-ground and all weather—really, all weather is the thing that we haven't had before. The radar allows you to see through cloud.

So when you have got a marine, a soldier, sailor, airman, or ally that is on the ground that needs fire support, bottom line, having the F–35 out there in numbers, either staged based afloat or based ashore, is going to allow us to provide those high-volume fires, aviation fires regardless of the threat. It is really an exciting time to be a young person coming in to fly Marine airplanes, Air Force airplanes, Navy airplanes. And then we do have to get the T–45 fixed. I was down there in Meridian, talking to those young sailors and marines, and they are chomping at the bit to get in these new airplanes.

Mr. KNIGHT. I always love to hear General Davis talk about the F–35. So it is a great advertisement. And I thought I would give you one last chance.

General Harris or General Bunch, the western ranges I always considered as part of kind of the infrastructure of what we do for air operations. And it is so important, especially with the F–35 and the F–22, because the ranges have to be so much bigger and more expansive. Of course, we can go out to the ocean and things of that nature, but give us an idea of what the western ranges are to air operations.

General BUNCH. Well, sir, the western test ranges are critical to our test and our training. They are critical to everything that we do. And we have a healthy investment into those to modernize them and make them more capable. We are also making—but as you said, given the fifth-gen planes and what they can do, we are now outgrowing some of our ranges to the point that we need more. So we are focusing in, also, on live virtual constructive to be able to create an environment where we can do those to the threat density we need and everything else.

But they are critical. We are—we do see them as a crown jewel. We are investing in them to modernize them, to make them more capable to do what we need, but we are also investing, as those fifth-gen capabilities grow, they are going farther and farther out, and we need to advance those capabilities as well.

Mr. KNIGHT. Thank you. Thanks, Mr. Chair.

Mr. TURNER. Mr. O'Halleran. And following Mr. O'Halleran, the next four are Kelly, Gaetz, Brown, and Bacon.

Mr. O'Halleran.

Mr. O'HALLERAN. I thank you, Mr. Chairman. And thank you to the witnesses for your service, your families' service, and for coming before us today.

My questions are for Admiral Grosklags and General Bunch. The budget request shows a drop in the production rate in 2018 for an air-to-air missile, the AIM–120D. That is a very high demand, and as we were told by the Chief of Staff of the Air Force and Chief of Naval Operations, a critical weapon Navy and Air Force need many more of right now. Specifically, the request is lowered by more than 300 missiles in comparison to projections in last year's budget.

Based on the budget, it appears the issue might be some kind of part supply—supplier or part obsolescence problem. Can you explain this reduction in the production rate? Is it due to a lack of funds or some other problem? If Congress were to add funding, could it be used to buy more missiles or are you limited by where things are with the production line? I understand wanting to limit your budget exposure under the assumption that the F3R [form, fit, function refresh] delay would limit fiscal year 2017 production quantities, but the redesign effort shows these reductions are premature when we really should be closing the inventory shortfall.

General BUNCH. Sir, I will take the first shot, then I will let G–8 jump in at the end of this.

So the driver for why the numbers do go down in the budget, we stay steady in 2017, and we do go down in 2018, in 2019 we go back up, it is tied to the difficulties we are having getting the form, fit, function refresh on schedule and progressing. It is something we continue to work with the company, but what we did was we intentionally laid in a line so that they had the ability to break that into the production without causing a more negative ramp. We actually have looked across our weapons inventories. As many know, we are using a lot of weapons right now. We are looking at what we need to build up. That is actually one that General Harris and I kind of agree, that is not one we need to try to add dollars to to do right now. We need them to get the form, fit, function properly done and get that on track.

Then once we get that on track, I would be willing to come back and ask you for additional help, but I am not ready to ask for that right now, sir.

Admiral GROSKLAGS. You know, from the Department of Navy's perspective, we are in exactly the same place. I think if you look at what we have budgeted for across the FYDP [Future Years Defense Program], which I know is not the issue here today, we have actually maintained the total quantity across the FYDP, but we did come down in FY 2018 and 2019 for the exact reasons that General Bunch stated.

Mr. O'HALLERAN. So just to clarify, there is an inventory shortfall on these weapons, but you have just decided to——

General BUNCH. We would love to buy more, sir, but I need them to get the production right before we buy too many more and I run out of parts on the line and I don't have the ability to produce.

Mr. O'HALLERAN. Thank you. And I yield.

Mr. TURNER. Mr. Kelly.

Mr. KELLY. Thank you, Mr. Chairman.

Vice Admiral, I understand the Navy's Next Generation Jammer Increment 1 program recently was able to accelerate some of the program's milestones throughout the life of the program. Can you tell us how important the updated jamming capability is to the warfighter, and are there additional efforts to get Increment 1 fielded faster? And I think you are the only vice—I don't do well with names so——

Admiral GROSKLAGS. That is okay. I will ask Admiral Miller to jump in if needed.

The program just recently passed its milestone B, so they are fully in their development stage at this point. They are on track. We anticipate a milestone C, which is lower then initial production in 2019 and fielding in 2021. This is one of former Secretary Kendall's accelerated programs, if you will, and we have been taking advantage of that at every turn. It has significantly less oversight within the building than most of our ACAT [Acquisition Category] I programs.

And the program, we are trying to accelerate it. We don't currently have any insurmountable technical hurdles, but there are some things we need to work through, particularly with the amount of radiated power we are trying to get out of that pod and the effects that it has on the rest of the aircraft systems when it is radiating.

So we plan to start flight tests late 2018, early 2019, and then we will get a full evaluation of those interference effects.

Mr. KELLY. And then just very briefly. I know that there are $2 billion in Increment 2, the budget shows $2 billion for Increment 2. Does this take into account any potential cost savings in the overall Increment 2 program if common hardware is used from Increment 1?

Admiral GROSKLAGS. It does, but right now we don't anticipate a significant amount of common hardware due to the significantly different frequencies we are dealing with and the location of the pod on the aircraft and, again, mutual interference effects.

Mr. KELLY. And then my next question deals with the MQ–4C Triton. In your testimony, you talk about it. In 2009, the Navy initiated demonstration program of unmanned maritime ISR using the older model Global Hawk that the Air Force was no longer using. We are still using it, and so—in support of the U.S. 5th Fleet in Central Command region. Can you tell us what capabilities the Triton will bring to the fleet and how they are shaped by what you learned from the demonstration?

Admiral GROSKLAGS. We continue to fly our BAMS [Broad Area Maritime Surveillance] demonstrator, the one aircraft that we have that was considered a 1- or 2-year demonstration program. I think we have had it out there flying in CENTCOM now for about 7 years. So we continue to gather data. The biggest changes for the Q–4 are in terms of its ability to fly in weather and turbulence conditions that some of its predecessors could not. So we did some significant structural modifications to that aircraft.

We are leveraging all the work that the Air Force has done ahead of us in terms of the TC [traffic control] pattern, the communications from that aircraft, the dissemination of the intelligence surveillance information. And as you know, we are quickly going to turn that after it IOCs into a multi-INT [multiple intelligence] sensor. So when we EOC [early operational capability] in 2018, right behind that in 2020 we will have an initial operational capability for that multi-INT sensor capability.

Mr. KELLY. Thank you, Mr. Chairman. I yield back.

Mr. TURNER. Mr. Gaetz.

Mr. GAETZ. Thank you, Mr. Chairman.

General Bunch, thank you for your service to the country and also to my district during your time at Eglin Air Force Base.

You are aware that the Gulf Range is important for the F–35 training mission. You are also aware that the equipment that supports the Gulf Range is very old. What importance does the Air Force place on upgrades to the Gulf Range so that we can be fully capable with the F–35?

General BUNCH. So in my previous position as a test center commander, we created what we needed to do to modernize. Now that path is coming back in and we are evaluating in our budget priorities. We have invested in some areas there, sir, to grow those capabilities. We know it is critical for what we need to be able to do to have the ranges that we need and the distance that we need to be able to execute things, and there is a lot of good things that can go in there that we are working on.

Mr. GAETZ. My next question is for Admiral Grosklags. I have a letter dated May 5th, it is signed by Chairman Turner, Ranking Member Tsongas, and myself. It requests the schedule and timeline for tests and evaluations of equipment associated with the T–45, including the equipment being tested, the location of those tests, the frequency of those tests, and how the results of the tests are informing the Navy moving forward. As of today, we have not received that information. Is that typical?

Admiral GROSKLAGS. No, sir, it is not. Typically, we try and respond within 10 days to 2 weeks. And I have to apologize for the fact that we haven't gotten that response to you. I know that it is in the process of being sent back over here to the Hill, but I can't give you the exact status or where it sits today.

Mr. GAETZ. I will yield my remaining minute and 40 seconds to the chairman so he can advise as to how I can get this information that is critically important to my constituents.

Mr. TURNER. I think you are doing it. As you know, we are working together on it, and it is great that you have once again raised this issue. The—I think we will leave this record open and also make any continuing obligation for an answer to this hearing. But, you know, clearly, this is a priority for the committee and Mr. Gaetz, and we want to see the request satisfied.

Admiral GROSKLAGS. Sir, understand.

Mr. TURNER. Mr. Brown.

Mr. BROWN. Thank you, Mr. Chairman. Over here? Yep.

This question is for Admiral Miller and General Harris. For the last 15 years, there has been a gap in the airborne ISR capabilities, or I should say capacity, and the demand or the need in the field. How do you go about determining what is an acceptable or optimal level of support? What is the aspirational goal? Where should we be? And what is the—in the 2½ minutes left, what is the plan to get us there?

Admiral MILLER. Well, earlier we talked about the Triton, MQ–4 Triton, and I will tell you, it is going to be a game changer. We have been flying the one BAMS demonstrator—the Broad Area Maritime Surveillance demonstrator, in 5th Fleet. Now we are talking about orbits worldwide with the Triton, with the extra sensors that it is going to provide, as well as the multi-INT capability that, as Admiral Grosklags said, will arrive later in 2020.

So what this will be able to provide to the theater commanders, to the fleet commanders, and contribute to basically all the forces that are in its AOR [area of responsibility] that it will be operating I think will be significant.

General HARRIS. Thank you, sir, that is a great question. We always try and provide everything to the warfighter, but the demand has been somewhat insatiable. So in our effort to make sure that we are providing a long-term capability, we have frozen our medium-altitude ISR at 60 lines with the intent of getting the team healthy so that we can surge if required for the future.

On that line, though, we continue to make improvements. And we are changing from the MQ–1 to an all MQ–9 fleet to give us better capability and capacity for both ISR and combat employment. In addition, in an unclassified format, upgrading the sensors on that particular weapon system. We also have high-altitude ISR that we continue to provide, and we are looking at recapping our big-wing ISR, which will help focus our medium-altitude ISR into more important areas and be on target much faster.

Mr. BROWN. Great. Unrelated follow-up, at Joint Base Andrews where I represent, the question was already asked about the radar system and, certainly, with that important mission homeland defense and respond to emergencies around the National Capital Region. We are looking forward to a timely delivery of those radars.

Thank you, Mr. Chairman. I yield back.

Mr. TURNER. Mr. Bacon.

Mr. BACON. Thank you. I want to thank all of you for being here today. And I want to recognize General Davis' service. You were great to work with in the current capacity, but also when you were in Cyber Command.

We have the best air power in the world, but it has been put under a lot of strain with continuing resolutions, you know, the BCA [Budget Control Act] caps. And we owe you better, and I hope we do that.

I have a couple questions on ISR and EW [electronic warfare]. First, a comment. I appreciate the fact that we are funding the U–2 program in conjunction with the Global Hawk. I think we need both. And there is a lot of unmet ISR needs out there, so I applaud that, I commend it.

A question with the E–8 program. I think we have a lot of unmet GMTI [Ground Moving Target Indicator] requirements out there, and I think we have a good plan for the recap or modernization. Do we have a shortage in funding, though, when it comes to the legacy E-8 as we transition? Because I am under the impression we do. I was just hoping you would comment.

General BUNCH. Sir, I don't believe we have a shortfall there in funding. What I have is a shortfall in the ability to get them in the air and have them available to support the mission. So the aircraft are old and the aircraft have diminishing—well, not so much diminishing, they are having higher downtime for heavy maintenance, and we have not been able to get as many of them in the air. It has taken longer to get them through the depot. So what we are looking at is how do we work with the contractor to do the depots in a more timely manner, to increase the number of aircraft that are available with this high-demand, low-density targets that we have.

Mr. BACON. Right.

General BUNCH. And the other item we are doing to make sure that we fill the gap until we get the recap effort going, we are doing service life studies in a variety of areas. We did one on the fuselage. That extended the life out more. And our intent is to ensure that we are doing everything we can to keep that aircraft viable until we get the recap done so we can meet the combatant commanders' requirements.

Mr. BACON. Okay. Thank you very much.

One last question, on the F-35, it has incredible sensors onboard. And I think we have a plan right now to do the links with the wingman, but I see there is a gap. I don't see where we are meeting that gap of getting that information off the F-35s back to home station while they are flying so that the follow-on missions will get that information faster.

Do we have a plan or a thought process how we are going to try to transition or get a plan where we can get this incredible data off the F-35 back more quickly so we can use it?

General DAVIS. I can take a crack at that, sir. We are doing some experiments with that right now. One, in conjunction with the Navy. The mission area——

Mr. TURNER. General, I am going to ask if you would submit the answer to that for the record. And we are going to take it that it is yes, but the detailed response, we are going to ask for you to submit to us, if you can submit that quickly.

[The information referred to was not available at the time of printing.]

Mr. TURNER. Let me go to Mr. Banks, with everybody's approval, for 2 minutes, and then we are going to close out with Ms. McSally, who is going to ask a question that the answer is going to be for the record.

So, Mr. Banks, you have 2 minutes.

Mr. BANKS. Thank you, Mr. Chairman.

Lieutenant General Harris, first of all, I commend you for your work in the President's budget to fully fund the A-10 weapon system. It has been and continues to be a great asset to our Air Force and is of significant importance to my district in northeast Indiana, which is the proud home of Indiana's Air Guard's 122nd Fighter Wing.

As you certainly know, over the past few years, this wing has been slated to convert their mission from the A-10 to the F-16. Yet, yesterday in the Senate, Secretary Wilson stated, quote, "The A-10 is at Fort Wayne, and we have no intention of removing it. It is there for the foreseeable future, and they will have that manned combat mission," end quote.

You stated in your testimony that the transition away from the system will occur in 2030. So my question is, how do you see that transition occurring for current squadrons, particularly those wings in our National Guard, and will they then utilize F-16s, with a transition directly from the A-10 to the F-35? How do you see that transition occurring?

General HARRIS. Well, we are committed to flying the A-10 weapon system and continuing to work through the upgrades to that as we are required to, and with the NDAA [National Defense Authorization Act] to compare it for the F-35 CAS [close air support] comparison task that we have.

Our intent is to fly that weapon system as long as we can, because we have made cuts in the A-10, the F-16, and pretty much all of our conventional fighters based on our budget. So if we had an unlimited budget, we would keep everything we have and just continue to grow an F-35 fleet.

At 48 to 60 F-35s a year, we will be retiring some of our older airplanes and putting newer airplanes or more capable airplanes into these units. Right now, the plan continues to be that, and we will address this, that as we reduce some of the A-10 capability that has been approved, it will be either with F-35s or with the F-16 for the weapon system.

Mr. BANKS. Thank you. I yield back.

Mr. TURNER. Gentlemen, many people have additional questions. They are going to submit those to you for the record. I am going to ask that you submit your answers within 10 days of receipt of them in writing. I have two members who are going to ask their questions for the record here now and not receive a response, but a written response from you. And it is Martha McSally, and then Mr. Langevin will be asking, and they are going to have 2 minutes.

Ms. MCSALLY. And as usual, thank you, Mr. Chairman. I have questions about the A-10. So in your testimony, you say you are committed to maintaining a minimum of six A-10 combat squadrons through 2030, with at least 171 combat-coded A-10s and 283 in the fleet, which we continually here protect every year. That is nine squadrons that we currently have.

The A-10s are now, as you know, on the DMZ [Demilitarized Zone] in South Korea. They are kicking butt against ISIS [Islamic State of Iraq and Syria]. They are deploying with the European Reassurance Initiative. I was over in Estonia, they are welcoming them to come back any time soon with the Russian aggression there.

From my view and my experience, if we need that capability until a proven tested replacement comes along, nine squadrons is the absolute minimum. So I really would like to know—I think this is the first time you guys have publicly said, although I have a pretty good intel [intelligence] sources with my buds in the A-10 community, but it is the first time you have publicly said that you are going to go down to six squadrons. I would really like to know what those planning assumptions are of the six squadrons that could be as soon as the testing and evaluation is done, so that could be in a few years from now. What are your planning assumptions? Are they 24 PAA [primary aircraft assigned] squadrons? Which squadrons are going away? Where is the A–X program, which we don't see in your testimony at all? How does this all fit together?

So again, a lot of questions on your assumptions here for this critical capability, and I look forward to hearing your responses on the record.

Thank you, Mr. Chairman.

[The information referred to can be found in the Appendix on page 116.]

Mr. TURNER. Last question, Mr. Langevin.

Mr. LANGEVIN. Thank you, Mr. Chairman. I want to thank all of our witnesses and thank you for your service.

General Harris, tactical data links are critical to mission success forming the core communication network that transmit the common operating picture, shared situational awareness, distributed sensor data for fusion, and integrated fire control across fighters, bombers, and surface ships. However, significant gaps remain in the tactical fighter networks that prevent the Air Force and other services from maximizing the combat utility of their aircraft, specifically, secure communication between the F–22 and the F–35 in an A2/AD [anti-access/area denial] environment.

And, General Pleus, the director of the F–35 integration, recently stated there is currently nothing on the books for any testing to solve this issue between fifth-generation aircraft. What is the Air Force doing to address the problem these two aircraft communicating during a penetrating strike or counter-air mission?

[The information referred to can be found in the Appendix on page 115.]

Mr. LANGEVIN. And then for all of our witnesses, while unmanned aircraft systems provide strategic ISR and combat capabilities, I also believe that these systems have the potential to be used for humanitarian operations in disaster areas abroad, particularly when it comes to mapping lightweight essential item delivery, damage assessment support, and increased situational awareness. How is the Department using unmanned aircraft systems in support of humanitarian missions abroad, and will modernization efforts support these disaster relief and humanitarian operations, and what more can be done to advance this concept?

[The information referred to can be found in the Appendix on page 115.]

Mr. TURNER. Thank you, Mr. Langevin. And if you could leave the written copy of that question so that they could have it behind so we don't have any delay in them receiving it.

I look forward to your written answers within 10 days. And as we stated before, we have votes on the House floor, so because of that, we will be adjourned.

[Whereupon, at 4:26 p.m., the subcommittee was adjourned.]

APPENDIX

June 7, 2017

PREPARED STATEMENTS SUBMITTED FOR THE RECORD

JUNE 7, 2017

Statement of the Honorable Michael Turner
Chairman, Subcommittee on Tactical Air and Land Forces
Navy, Marine Corps and Air Force Combat Aviation Programs
June 7, 2017

The hearing will come to order.

The subcommittee meets today to review the Navy, Marine Corps, and Air Force aviation investment and modernization budget requests for fiscal year 2018.

I would like to welcome our distinguished panel of witnesses:

- Vice Admiral Paul Grosklags, Commander of the Naval Air Systems Command;
- Rear Admiral "Chip" Miller, Director of the Navy's Air Warfare Division;
- Lieutenant General Arnold Bunch, Military Deputy in the Office of the Assistant Secretary of the Air Force for Acquisition;
- Lieutenant General Jerry Harris, Air Force Deputy Chief of Staff for Plans, Programs, and Requirements; and,
- Lieutenant General Jon Davis, Deputy Commandant of the Marine Corps for Aviation.

I understand this will be General Davis's final appearance before the committee because he will be retiring next month. General Davis, we thank you for your 37 years of distinguished service to the Marine Corps and our Nation, and wish you the best in your future plans.

I thank all of you for your service and look forward to your testimony today.

As I have stated at previous hearings, I support the President's commitment to rebuilding the capacity and capability of our military. However, I am concerned that the current budget request of $603 billion for the Department of Defense will not achieve that goal in the timeline desired and needed.

For example, at Congress' request, the military services submitted their unfunded requirements lists to the congressional defense committees last week. The total amount was over $30.0 billion dollars. A significant portion of these requests were related to modernization needs.

Of particular note, the Air Force included an additional 14 F-35A aircraft, and the Navy and Marine Corps included 20 additional aircraft comprised of F-18 Super Hornets, F-35Bs and F-35Cs.

I suspect all of our witnesses today will support the President's budget request. However, Members of this subcommittee need to better understand what additional capabilities are required above the President's budget request, and why it is imperative we work to fully resource these unfunded requirements to accelerate the restoration of full-spectrum readiness.

I continue to support Chairman Thornberry and Chairman McCain who believe that a $640 billion dollar budget in fiscal year 2018 is required to build the capability needed for today's complex and dangerous world. The military services unfunded requirements also validate a higher topline funding level.

Today the subcommittee will review a broad portfolio of tactical aviation modernization programs and associated acquisition strategies.

The witnesses have been asked to identify their top five modernization requirements and briefly summarize how the budget request addresses them.

The aviation budgets for the Navy, Marine Corps and Air Force appear to be placing a higher priority on current readiness—and rightfully so.

Earlier this year, we heard the Vice Chief of Naval Operations report that over 60 percent of F-18s are out of service due to backlogs in depot repair. This budget fully funds depot capacity. This budget also robustly funds preferred aviation munitions such as the Joint Direct Attack Munition, Small Diameter Bombs, and Hellfire missiles—a much needed increase especially for the on-going combat operations in the U.S. Central Command's area of responsibility.

However, the Navy continues to absorb significant risk in its management of the strike fighter inventory. The Navy is challenged to replace legacy F-18C, D, and AV-8B aircraft that have reached the end of their service-life before they can be replaced by new F-35s or new F-18 Super Hornet aircraft. Each year, the Navy flies about 180,000 flying hours in its F-18 fleet which equates to the entire fleet expending 24 to 36 aircraft worth of service-life per year.

The Navy's fiscal year 2018 budget request includes the procurement of 38 F-18 Super Hornets and F-35s, so the Navy is only slightly above its annual expenditure of fighter aircraft life. There are two less Navy F-35Cs in the budget request than were projected last year, and the future years defense program for fiscal year 2018 reduces the planned F-35C procurement by seven aircraft.

The Air Force currently has a fleet of 55 combat-coded fighter squadrons, significantly smaller than the Desert Storm force of 134 fighter squadrons.

We have heard testimony from senior Air Force leaders that with the current demand for rotational fighter presence, 55 combat fighter squadrons does not allow sufficient time to train pilots or maintain aircraft, which contributes to degraded full spectrum readiness.

The 55 combat squadron level meets the minimum requirements set forth in current Defense Planning Guidance, but the Air Force considers it to be high risk in more challenging scenarios.

The Air Force has stated that in order to meet current steady-state demands and maintain readiness to meet most surge requirements, the Air Force needs to grow to at least 60 combat fighter squadrons, invest in munitions, modernize existing platforms, and increase participation in advanced training opportunities.

The fiscal year 2018 budget request includes 46 F-35As for the Air Force and that number is two more than planned last year for this budget request. However, I would also note that two years ago, the Air Force planned to procure 60 F-35As in fiscal year 2018. In testimony before this committee last July, Air Force General "Hawk" Carlisle, the former Commander of Air Combat Command, testified that to address the Air Force's capability and capacity shortfalls, the desired production rate is 60 F-35s per year.

The committee is also pleased to see the Air Force reverse its decision to retire the U-2 in 2019 and provide funding to maintain both the U-2 and Global Hawk platforms to meet high-altitude airborne intelligence, surveillance, and reconnaissance demands of our combatant commanders. However, all of the services represented today need to take a hard look at their investment strategies in airborne ISR capacity because a significant portion of combatant commander's requirements are still unmet in many of the intelligence disciplines.

As the committee continues its deliberation on the fiscal year 2018 budget request, we will look for opportunities to further address the Services' most pressing modernization requirements.

Without objection, all witness' prepared statements will be included in the hearing record.

Admiral Grosklags please proceed, followed by General Davis, Admiral Miller, General Bunch and General Harris.

NOT FOR PUBLICATION UNTIL RELEASED BY
THE HOUSE ARMED SERVICES COMMITTEE
TACTICAL AIR AND LAND FORCES
SUBCOMMITTEE

STATEMENT OF

VADM PAUL GROSKLAGS
REPRESENTING ASSISTANT SECRETARY OF THE NAVY
(RESEARCH, DEVELOPMENT AND ACQUISITION)

AND

LIEUTENANT GENERAL JON DAVIS
DEPUTY COMMANDANT FOR AVIATION

AND

REAR ADMIRAL DEWOLFE H. MILLER III
DIRECTOR AIR WARFARE

BEFORE THE

TACTICAL AIR AND LAND FORCES SUBCOMMITTEE

OF THE

HOUSE ARMED SERVICES COMMITTEE

ON

DEPARTMENT OF THE NAVY'S AVIATION PROGRAMS

June 7, 2017

NOT FOR PUBLICATION UNTIL RELEASED BY
THE HOUSE ARMED SERVICES COMMITTEE
TACTICAL AIR AND LAND FORCES SUBCOMMITTEE

INTRODUCTION

Mr. Chairman, Representative Tsongas, and distinguished members of the Subcommittee, we thank you for the opportunity to appear before you today to discuss the Department of the Navy's (DoN) Aviation programs. Our testimony will provide background and rationale for the Department's Fiscal Year 2018 aviation programs budget request aligned to our strategic priorities and budgetary goals.

The United States is a maritime nation with global responsibilities. Our Navy and Marine Corps' persistent presence and multi-mission capability represent U.S. influence across the global commons. We are an agile maritime strike, amphibious and expeditionary power projection force in readiness, and such agility requires that the aviation arm of our naval strike and expeditionary forces remain strong. Our budget submission ensures Naval Aviation possesses the capability, capacity and readiness to deliver five essential functions outlined in our maritime strategy – All Domain Access, Deterrence, Sea Control, Power Projection and Maritime Security. These key, essential functions are missions that depend upon Naval Aviation to enable their success.

Global connections continue to multiply, fueled by rapid advances and proliferation of technology, particularly cyber and other information technologies. Our competitors are pursuing advanced weapon systems at a development pace we have not seen since the mid-1980s, and because of these focused pursuits; both near-peer nations and non-state actors pose credible threats to our security. As such, it is imperative that we fund a force with the capability and capacity to fight and win against any of our five major challengers (China, Iran, North Korea, Russia, and Violent Extremism) by investing in advanced systems that increase lethality for both the current and future force.

Our ability to respond to this dynamic strategic environment, high operational tempo and evolving Combatant Commander (CCDR) requirements continues to be constrained by current fiscal realities. The Department is still recovering from appropriations that were significantly lower than the budget requests for Fiscal Years 2013-2016. We strive to improve efficiency in our own internal business practices to make every dollar count, but our efforts are undermined by the absence of stable, timely budgets and resources aligned to stay ahead of the threats. We encourage Congress to re-evaluate the Budget Control Act caps, as outlined by our President's Budget request. Timely passage of a full year appropriation as at requested level will provide for the most efficient execution of the resources provided by Congress, while bringing stability to our workforce and the industrial base.

This fiscal context drives difficult choices to best balance between capability, capacity, readiness and the industrial base. Our Fiscal Year 2018 budget supports the five essential functions outlined in our maritime strategy, the operational context we as a Nation find ourselves in, and the current fiscal environment.

Our investments are focused, balanced and prioritized to deliver and support a global sea-based and expeditionary force. Our budget is based on the transition of major components of the Carrier Air Wing (CVW), Expeditionary Strike Group and land-based Expeditionary Wings, and includes: manned and unmanned aviation system teaming; integration of warfighting capabilities to ensure multiple systems operate together across platforms, weapons, networks and sensors; advanced computing; and incorporation of commercially driven technology and additive manufacturing to provide a technological advantage over adversaries.

The Department continues to pursue acquisition and business process reform measures to deliver capability faster at reduced cost. New measures include implementation of accelerated acquisition policies for Rapid Prototyping, Experimentation and Demonstration; establishment of Maritime Accelerated Capability Office programs; and the use of Rapid Deployment Capability processes. We are actively promoting innovation and the transition of key manufacturing technologies and processes with investments focused on affordability and those most beneficial to the warfighter. There is also a continuing transition from a hardware-centric world to a software-centric world by leveraging common development standards and requirements for modular weapon system components.

The Navy/Marine Corps "Vision for Naval Aviation 2025" provides the framework for determining investment priorities across the triad of warfighting capability, capacity, and Naval Aviation wholeness. There are several central themes to our 2018 Naval Aviation budget plan: next generation fighter/attack capability; unmanned systems; netted persistent multi-role intelligence, surveillance, reconnaissance (ISR) and targeting; advanced strike and networked enabled weapons programs; supporting capabilities such as electronic attack and electromagnetic spectrum superiority, maritime patrol, and vertical lift; readiness; and targeted modernization of the force for relevance and sustainability.

The best way for pilots to train for combat is by flying their aircraft in live, scenario-based training missions against live opposition. However, many elements of combat cannot be replicated in the training environment. The Department is committed to augment aircraft flight hours by providing high-end virtual training. To do that, we are making investments in Live, Virtual, and Constructive Training that enable our aircrews to link across the country and train in high fidelity simulators. As we develop these technologies, the Department plans to connect aircrews in live flying aircraft against synthetic adversaries. We are also dedicated to leveraging the Science of Learning into all levels of aviation training. To do this, we are exploring innovative ways to leverage big data/analytics and flexible training systems that will maintain the nation's competitive advantage.

At its foundation, as core unpinning principals, Naval Aviation is actively pursuing and seizing innovation and advantage wherever it can. We are implementing our vision toward greater tactical and technical innovation to provide the right capability in the hands of the warfighter, on schedule, and in the most affordable manner possible.

TACTICAL AVIATION

F/A-18 Overview

There are four Navy and eleven Marine Corps F/A-18A-D active strike fighter Hornet squadrons with a total inventory of 546 aircraft. There are 30 Navy Super-Hornet (F/A-18E/F) strike fighter squadrons with a total inventory of 544 aircraft. Combined,

F/A-18A-D Hornets and F/A-18E/F Super-Hornets have conducted more than 219,454 combat missions since September 11, 2001.

F/A-18 A/B/C/D Hornet

Based on inventory modeling, a portion of the existing inventory of 546 Navy and Marine Corps F/A-18 A-D aircraft will be flown through the mid-2030 timeframe. The DoN will continue to meet Navy active F/A-18A-D squadron operational commitments until 2027, Marine Corps active and reserve squadrons until 2030, and Navy reserve squadrons through 2034.

To support this Fleet plan, the Fiscal Year 2018 President's Budget requests $294 million in APN to implement aircraft commonality programs, enhance relevant capability, improve reliability, and ensure structural safety of the inventory of 546 F/A-18 A-D Hornets; $31.4 million is for a Service Life Extension Program (SLEP). The funding priorities for F/A-18 A-D Hornet will be safety, reliability, and capability.

Service life management efforts have extended the F-A-18 A-D beyond its original service life of 6,000 flight hours to 8,000 flight hours with select aircraft that may be extended up to 10,000 flight hours. Discovery of unanticipated corrosion on these legacy jets complicates depot throughput, and service life extensions for aircraft with more than 8,000 flight hours require High Flight Hour inspections, which furthers increases maintenance-man hours. These inspections assess the material condition of each aircraft and apply a unique combination of inspections and airframe modifications to maintain airworthiness certification. As of April 2017, 92 percent of the F/A-18 A-D fleet has over 6,000 flight hours and 24 percent have flown more than 8,000 flight hours; the highest flight hour airframe has attained over 9,799 hours. The Department endeavors to return the required number of aircraft to the flight line with the necessary capability upgrades, but remains concerned that low reliability rates will affect our ability to train and fight.

In addition to the flight hour extension strategy, these aircraft require capability upgrades in order to maintain warfighting relevancy. The Department will continue to procure and install advanced systems such as mission computers, aircraft survivability equipment, radios, radars and targeting pods on select F/A-18 A-D aircraft to counter evolving threats. However, while the DoN continues investing in warfighting upgrades in order to maintain tactically relevant aircraft, the Services are challenged to improve the reliability of this aged airframe.

F/A-18E/F Super Hornet

The F/A-18E/F Super Hornet will be the numerically predominant aircraft in the Navy's CVW Strike Fighter force through 2035. The F/A-18E/F began Full Rate Production (FRP) in 2000. Continued investment in capability upgrades significantly improves the lethality of the CVW.

The Fiscal Year 2018 President's Budget requests $1.25 billion in APN for 14 F/A-18E/F Super Hornet aircraft and $251.2 million of RDT&E for F/A-18 Block III, IRST Block II, F/A-18E/F Service Life Assessment Program (SLAP), radar upgrades and improvements. With the support of Congress, we will also procure a minimum of 80 additional Super Hornets across the Future Years Defense Program (FYDP) and continue modernization plans to address continuing warfighter demand for advanced tactical aircraft. These additional procurements begin to mitigate the decline in DoN's strike fighter inventory and enable older aircraft to be pulled from service for mid-life upgrades and rework to extend their service life.

The Super Hornet modernization plan features an incremental approach to add conformal fuel tanks to extend range and replace outdated electronics. Other modernization efforts will incorporate new technologies and capabilities, to include, Digital Communication System Radios, Advanced Targeting Forward Looking Infrared (with shared real-time video), Accurate Navigation Distributed Targeting System, Infrared Search and Track, and continued advancement of the APG-79 Active Electronically Scanned Array Radar.

Due to high utilization rates, the F/A-18E/F fleet has flown approximately 47 percent of the total flight hours available within the 6,000 hour limit design life. The remaining fleet flight hour capacity will be inadequate to meet operational commitments starting in the early 2020s. As a result, the Department initiated a phased F/A-18E/F SLAP to determine requirements to extend the airframe service life beyond 6,000 flight hours. The F/A-18E/F SLAP incorporates lessons learned from the F/A-18A-D SLAP and SLEP analysis and was initiated earlier in the F/A-18E/F life-cycle. Super Hornet SLAP commenced in 2008 with completion expected in 2018. The SLAP goal is to analyze actual Fleet usage versus structural test data to support the design of Service Life Modifications (SLM) that will ultimately extend F/A-18E/F service life from 6,000 to 9,000 flight hours. The initial phases of the F/A-18E/F SLM began in 2014 with the development and fielding of Engineering Change Proposal kits to upgrade life-limited areas of the F/A-18E/F that were revealed by SLAP analysis.

EA-18G Growler

The EA-18G Growler is a critical enabler for the Joint force. EA-18G brings fully netted warfare capabilities to the fight, providing unmatched agility in the Electromagnetic Maneuver Warfare environment. The Fiscal Year 2018 President's Budget requests $173.5 million of RDT&E for modernization.

To date, 136 EA-18G aircraft have been delivered, representing 85 percent of the funded inventory objective. Initial Operational Capability (IOC) occurred in September 2009 and the Fleet Response Plan was approved in November 2009. Since their initial deployment, Growlers have flown more than 2,300 combat missions and have expended approximately 16 percent of the 7,500 flight hour life per aircraft. Electronic attack capabilities, both carrier-based and expeditionary, continue to mature with development of the Next Generation Jammer (NGJ). NGJ Increment 1 is scheduled to begin replacing the legacy ALQ-99 Tactical Jamming System in Fiscal Year 2021. Additionally, we continue to invest in the EA-18G passive detection and identification capabilities while improving network connectivity to provide overall battlespace awareness and targeting for the carrier strike group.

The recent authorization of seven additional EA-18Gs will extend aircraft deliveries into Fiscal Year 2018. With the seven additional aircraft, the total procurement quantity of 160 EA-18Gs fulfills Navy requirements for carrier-based Airborne Electronic Attack (AEA) and expeditionary EA-18G squadrons.

Additional EA-18Gs, above the funded procurement objective of 160, may be explored by the Department of Defense as it considers options to support an AEA force structure that meets the Joint Warfighter requirement.

AV-8B Harrier

Since the beginning of the war on terror, the AV-8B Harrier has been a critical part of the strike fighter inventory for the Joint force. This aircraft has flown more than 60,000 hours in combat since 2003, an average of over 475 hours per aircraft, with zero losses from the enemy in the air, but six losses on the ground when the enemy broke through our perimeter at Bastion Air Base in 2012.

The Fiscal Year 2018 President's Budget requests $48.8 million in RDT&E funds to continue Design, Development, Integration and Test of various platform improvements. These improvements include an Engine Life Management Program, Escape Systems, Joint Mission Planning System updates, Link 16 Digital Interoperability (DI) integration, Operational Flight Program (OFP) block upgrades to mission and communication systems, navigation equipment, weapons carriage, countermeasures, and the Obsolescence Replacement/Readiness Management Plan. Additionally, the Department is requesting $43.6 million in APN funds to continue the incorporation of Obsolescence Replacement/Readiness Management Plan systems, electrical and structural enhancements, inventory sustainment and upgrade efforts to offset obsolescence and attrition, LITENING Pod upgrades, F402-RR-408 engine safety and operational changes, and DI upgrades that include Link 16.

The AV-8B continues to deploy in support of operational contingencies and each Marine Expeditionary Unit (MEU) deploys with embarked AV-8Bs. The AV-8B equipped with LITENING targeting pods and a video downlink to ROVER ground stations, precision strike weapons, Intrepid Tiger II Electronic Warfare (EW) pods and beyond visual range air-to-air radar guided missiles, continues to be a proven, invaluable asset for the Marine Air Ground Task Force (MAGTF) and Joint commander across the spectrum of operations. AV-8B squadrons, both land- and sea-based, have flown more than 10,000 hours of strike sorties against Islamic State in Iraq and Syria with an average combat radius of 900 miles. Digital Improved Triple Ejector Racks have allowed us to load up to six precision guided munitions per aircraft, with fuel tanks, guns, and LITENING Pods, exponentially increasing the combat viability of this platform. Airborne Variable Message Format terminals are currently being installed in AV-8B to replace the current digital-aided Close Air Support (CAS) technology. The program will continue development of the H6.2 OFP which includes initial integration of Link 16 message sets. Additionally, this OFP will integrate Federal Aviation Administration compliant Navigation Performance/Area Navigation capability, an update to the LITENING Common OFP to implement improvements to moving target tracking, and correction of software deficiencies identified through combat operations. The program will also work on the H7.0 OFP which will integrate full Link 16 functionality. As an out-of-production aircraft, the AV-8B program continues to focus on sustainment efforts to mitigate significant inventory shortfalls, maintain airframe integrity, achieve full Fatigue Life Expended, and address reliability and obsolescence issues of avionics and subsystems.

Operations ODYSSEY DAWN, ODYSSEY LIGHTNING, ENDURING FREEDOM, FREEDOM SENTINEL, and today's Operation INHERENT RESOLVE confirm the expeditionary advantages of Short Take-Off and Vertical landing (STOVL) capabilities. Placing the Harrier as the closest multi-role fixed-wing asset to the battlefield greatly reduces transit times to the fight and enables persistent CAS without strategic tanking assets. Airframe sustainment initiatives, capability upgrades, and obsolescence mitigation is essential and must be funded to ensure the AV-8B remains lethal and relevant.

F-35 Lightning II

The F-35 Lightning II will form the backbone of U.S. air combat superiority for decades to come. The F-35 brings unprecedented low observable technology, modern weaponry, and electronic warfare capability to the Navy and Marine Corps. Delivering this transformational capability to front-line forces as soon as possible remains a top priority. The F-35 will replace legacy tactical fighter fleets of the Navy and Marine Corps with a dominant, multirole, fifth-generation aircraft, capable of projecting U.S. power and deterring potential adversaries. The Fiscal Year 2018 President's Budget requests $550 million in RDT&E to support system design and development close-out and ramp-up Follow-on Modernization and $3.9 billion in APN for 20 F-35B aircraft, 4 F-35C aircraft, modifications and spares.

The F-35 has flown over 70,000 flight hours, including approximately 27,000 for the F-35B and 7,000 hours for the F-35C. Marine Fighter Attack Squadron (VMFA) 121, the first IOC squadron, is now forward deployed in Japan defending the Nation's interests abroad. In 2018, the Navy and Marine Corps team will deploy two MEUs with a detachment of F-35Bs aboard ship marking the first extended at sea deployments for F-35. The fielding of the Marine Corps' F-35B STOVL variant continues to make excellent progress due to the combined efforts of the Department, industry, and Congress. Critical Military Construction (MILCON) at our bases and air stations is underway both at home and overseas to support this fifth generation capability. Due to the level of effort, funding, and timely MILCON, the Marine Corps' transition plan remains on-track. VMFA-211 stood up in July 2016 on Marine Corps Air Station, Yuma, AZ and the Marine Corps' will transition its third operational squadron, VMFA-122, to the F-35B in 2018.

The F-35B achieved a number of operational and training milestones. Operationally, the Marine Corps has permanently stationed an F-35B squadron in Japan, conducted trans-oceanic flights across both the Atlantic and Pacific, and exercised the expeditionary capability of the aircraft both aboard ship and in austere environments. In training, Marine Corps has seen return on training investments. The first two F-35B pilots graduated flight school and have conducted sustained training operations across the range of military operations, including participation in large-scale joint exercises like "Red Flag". Pilots and instructors continue to praise the F-35 situational awareness and lethality as it achieves mission success previously unrealized in legacy platforms.

The Navy's first F-35C squadron begins transition in 2018. Navy IOC is event-driven and expected in the late 2018 to early 2019 timeframe. The first F-35C aircraft carrier deployment is planned for 2021. The Marines begin their first F-35C squadron transition, VMFA-314, in 2018, will be ready for expeditionary operations by 2020 and deploy aboard a carrier in 2022. Together, the Navy and Marine Corps will be operational in 2020 and replace our aging aircraft inventory with the greatest practical speed. The F-35B/F-35C aircraft will help recapitalize some of our oldest aircraft – our legacy F/A-18s – which are rapidly approaching the end of their service lives.

F-35 employs a block upgrade program to usher in new and advanced war-fighting capabilities. Whether the mission requires the execution of strike, CAS, counter air, escort, or EW, this aircraft is the key to our future. It empowers our maritime forces to fight from sea bases and expeditionary bases ashore anywhere in the world. However, to take full advantage of the aircraft's advanced capabilities and to keep the transition from legacy platforms on-track, this effort requires the continuation of the support received from Congress thus far.

The F-35 continues to mature and progress with programs in development and design, flight test, production, fielding, base stand-up, sustainment of fielded aircraft and stand-up of a global sustainment enterprise. The final system development and demonstration configuration, Block 3F, is finishing its final developmental test flights and our overall assessment is that steady progress continues to be made on all aspects of the program, although not without risk in software development and integration. This risk will continue to decline as the Department learns and makes adjustments. The discipline instilled several years ago in the method by which software is developed, lab tested, flight tested, measured and controlled has resulted in improved and more predictable outcomes.

The program has delivered over 230 aircraft to test, operational, and training sites, with the production line delivering F-35s on schedule. It remains a clear and prominent priority for the Department to complete the development phase on cost and schedule. DoN is committed to drive aircraft production cost and life-cycle costs down. As examples of cost reduction efforts, combined government and industry teaming has reduced aircraft production costs through "blueprint for affordability" initiatives and reduced F135 engine costs through ongoing engine "war on cost" strategies.

These affordability efforts include up-front contractor investments in cost reduction initiatives that are mutually agreed upon by the government and contractor team. This arrangement motivates contractors to accrue savings as quickly as possible in order to recoup their investment, and benefits the government by realizing cost savings at the time of contract award. The Department's goal is to reduce the flyaway cost of the USAF F-35A to between $80 and $85 million dollars by 2019, which is anticipated to also decrease the cost to the Marine Corps F-35B and Navy F-35C variants. The Department set a goal of decreasing overall operating and support life-cycle cost by 30 percent.

Next Generation Air Dominance (NGAD) Family of Systems

The Department initiated a Next Generation Air Dominance (NGAD) Analysis of Alternatives (AoA) in January 2016 to address the anticipated retirement of the F/A-18E/F and EA-18G aircraft beginning in late 2020 early 2030 timeframe.

The Joint Chiefs of Staff approved the Initial Capabilities Document that frames NGAD study requirements to support the full range of military operations from carrier-based platforms. The AoA will consider the widest possible range of materiel concepts while balancing capability, cost/affordability, schedule, and supportability. It will assess manned, unmanned, and optionally manned approaches to fulfill predicted 2030+ mission requirements. Analyses will consider baseline programs of record (current platforms), evolutionary or incremental upgrades to baseline programs (including derivative platforms), and new development systems or aircraft to meet identified gaps in required capability.

STRIKE FIGHTER INVENTORY MANAGEMENT

Through 2009, the Department's Strike Fighter force was relatively healthy. Several events transpired since 2009, however, which drove our current Strike Fighter inventory shortfall. The Budget Control Act of 2011 started multiple years of reduced military funding and F-35B/C fielding plans were delayed. As a result, the DoN decided to extend the life of legacy F/A-18A-Ds using our aviation depots. Sequestration led to furlough and a hiring freeze of a skilled government civilian artisan workforce at aviation depots, significantly impacting depot throughput and fleet readiness along with other factors such as high utilization rates, lack of aircraft procurement and lack of spare parts. Throughout this period, the operational demand for Naval Aviation forces remained high and accelerated the consumption of existing fleet aircraft. In essence, consumption of aircraft exceeded the new and rework production capacity of aircraft and caused an increasing shortfall.

The Naval Aviation Enterprise (NAE) aggressively tackled Strike Fighter Inventory Management (SFIM) to ensure that deployed forces are properly manned, trained and equipped. Each budget year, the NAE attempts to harmonize available funding between flying hours and readiness enabler accounts in order to achieve the greatest return on investment towards improved readiness.

Under the current budget and with Secretary Mattis' focus on readiness, aviation spares and readiness enabler accounts are receiving improved funding levels. It is important to note, however, that years of underfunding cannot be corrected in one budget year and will require stable, predictable funding over multiple years to achieve positive results. This shortfall will take time and likely require several years to correct.

The DoN has accepted significant risk in SFIM. The Department remains challenged with planning for F/A-18A-D and AV-8B aircraft that reach the end of their service life before replacement aircraft (F-35B/C or follow on F/A series) can be delivered into service. Fiscal Year 2018 investments begin to address the gap between the Strike Fighter inventory forecasts and Global Force Management Allocation Plan (GFMAP) demands by fully funding depot capacity. Near-to-mid-term risk remains due to uncertainty in readiness accounts and procurement levels that fail to match Strike Fighter service life consumption. Mid-to-long-term risk is driven by a shortfall in tactically relevant aircraft to replace F/A-18E/Fs that are soon to be inducted into commercial depots for SLM. Long-term risk is driven by Strike Fighter procurement that fails to match Strike Fighter service life consumption and attrition.

SFIM should be viewed in two separate and distinct phases. The near-term challenge is managing a DoN Tactical Aviation (TACAIR) force that has been reduced in capacity through a combination of historically high TACAIR utilization rates, constrained resourcing of sustainment and enabler accounts resulting in inadequate availability of spare parts, F/A-18 depot production falling short of the required output, and reduced Strike Fighter aircraft procurement. TACAIR aviation depots are expected to continue to improve productivity through 2019. In 2019, the focus will shift toward F-35 repair and begin to support F/A-18E/F SLM. In a similar effort to increase Harrier aircraft availability, the Marine Corps conducted a Harrier Independent Readiness Review which identified a need for changes in the Harrier sustainment plan to achieve required flight line and inventory readiness. This year, with sufficient resources, the Department is implementing these changes to return Harrier readiness to the required T 2.0 levels.

In the far-term, Strike Fighter inventory is predominantly affected by new procurement of F-35B/Cs and F/A-18E/Fs, as well as the F/A-18E/F SLM of our current fleet. CCDR driven operational demand, Fleet Response Training Plans and readiness requirements are expected to continue to drive increased Strike Fighter utilization rates that outpace procurements.

The DoN program of record is 680 F-35 aircraft. The Navy F-35C requirement is 340 aircraft, which includes 67 Marine Corps F-35C aircraft. Due to evolving circumstances, the total Marine Corps F-35 requirement is 420 aircraft; 353 F-35Bs and the 67 F-35Cs. The Navy and Marine Corps will continue to modify transition plans to take advantage of any possible F-35 accelerated procurement. Due to delays in the F-35 program and a changing threat environment, sustainment and modernization funding will be required to maintain the relevant operational capability of the F/A-18A-F and the AV-8B.

Strike-Fighter Force Structure

The 1,174 aircraft Strike Fighter force provides the projected DoN inventory needed to support the anticipated operational demand of nine CVWs through the 2025 timeframe. The Navy inventory requirement of 779 aircraft supports 36 active duty DoN Strike Fighter squadrons (with a mix of 10-12 aircraft per squadron). This requirement includes four Marine Corps Strike Fighter squadrons and is composed of 396 aircraft and two reserve squadrons with 22 total aircraft assigned. In order to maintain the operational aircraft, support aircraft are required for aviator training, flight-test, attrition reserve and the depot pipeline. This inventory entitlement is estimated based on historical averages and supports the validated requirement of four Strike Fighter squadrons per CVW. Through detailed analysis, inspections and structural repairs, the DoN has been successful in extending F/A-18 A-D aircraft to 8,000 flight hours - 2,000 flight hours beyond the original designed service life. Future inventory projections are based on a service life extension for F/A-18E/F aircraft to 9,000 flight hours from the current design life of 6,000 flight hours.

The Department's F-35C Strike Fighter program requires 14 active Navy squadrons, four active Marine Corps squadrons, and two training squadrons. The F/A-18E/F capabilities complement the F-35C and enhance the overall carrier-based warfighting capabilities. This force structure supports the operational demand per the GFMAP and projected aircraft carrier deployments. The Marine Corps' F-35B Strike Fighter program requires 14 active, 2 reserve and 2 training squadrons. Integral to DoN's current force structure reductions, tactical aviation squadrons were restructured to optimize the support they provide to the MAGTF and the Joint force.

PHYSIOLOGICAL EPISODES

The status of DoN efforts to address Physiological Episodes can be found at Addendum A.

AIRBORNE ELECTONIC ATTACK (AEA)

Next Generation Jammer (NGJ)

The Next Generation Jammer (NGJ) is the follow-on to the Vietnam-era AN/ALQ-99 initially fielded in 1971. The ALQ-99 has reached its capability limit both technologically and materially and is challenged against modern state-of-the-art computerized surface-to-air missiles systems. NGJ is designed to provide improved capability in support of Joint and coalition air, land and sea tactical strike missions and is critical to Navy's vision for the future of strike warfare. It will be DoDs only comprehensive tactical airborne electronic attack platform and is required to meet current and emerging threats. NGJ will use Active Electronically Scanned Array technology to provide full-spectrum dominance, the ability to jam multiple frequencies at the same time, higher radiated power, increased precision, and the application of digital techniques to counter increasingly advanced and sophisticated adversary radars and communications systems. NGJ will be implemented in three increments: Mid-Band (Increment 1), Low-Band (Increment 2), and High-Band (Increment 3).

Our Fiscal Year 2018 budget request of $632.9 million RDT&E,N is vital to maintain Increment 1 schedule, continue procurement and assembly of the Engineering and Development Models, and commence developmental flight testing. In addition, $66.7 million RDT&E,N is requested to complete Increment 2 technology feasibility studies and initiate technology demonstration efforts.

MAGTF Electronic Warfare/EA-6B Prowler

The Fiscal Year 2018 President's Budget request includes $29.6 million in RDT&E,N and $10.1 million in APN for MAGTF EW.

The MAGTF EW approach to Electromagnetic Spectrum Operations (EMSO) is a distributed, platform-agnostic strategy where every platform contributes and functions as a sensor, shooter and sharer to include EW. Marine Aviation is integrating EW systems and Intrepid Tiger II (IT II) payloads across all aviation platforms to provide commanders with an organic and persistent airborne EW capability - for every MAGTF - large and small. Included in this plan are the IT II EW payload, the F-35s organic EW capabilities, and the EW Services Architecture network to facilitate collaborative EW Battle Management.

IT II is a precision EW system providing EW Support and Electronic Attack capabilities. IT II has been integrated on the AV-8B, F/A-18A-D, and UH-1Y. Since 2012 IT II has completed over 20 deployments, and is currently deployed with the 11th, 24th, and 31st MEUs. Future aviation platforms for IT II integration are the MV-22B, KC-130J, AH-1Z, CH-53K, and RQ-21. Development of an IT II counter-radar capability began in Fiscal Year 2016 and will be fielded on the AV-8B, F/A-18A-D, and MV-22B from Fiscal Years 2020-2022. The F-35 brings a powerful combination of EW, weapons, sensors, and reduced signature to the MAGTF.

Currently, there are 18 EA-6Bs distributed to two Marine Corps operational squadrons, one deactivating Marine Corps squadron, and one Navy flight test squadron. Final retirement of the EA-6B from the DoN inventory will be in Fiscal Year 2019.

Future aviation EW capabilities will also be provided by the MAGTF Expeditionary Unmanned Aviation System (MUX). In addition to providing persistent reconnaissance, surveillance and communications, MUX will also provide a long range, persistent, penetrating and responsive airborne EMSO capability.

OTHER ELECTRONIC WARFARE INQUIRIES

Responses to Congressional requests for updates on electronic warfare can be found at Addendum B.

AIRBORNE EARLY WARNING AIRCRAFT

E-2D Advanced Hawkeye (AHE)

The E-2D AHE is the Navy's carrier-based Airborne Early Warning and Battle Management Command and Control system. The E-2D AHE provides Theater Air and Missile Defense and is capable of synthesizing information from multiple onboard and off-board sensors, making complex tactical decisions and then disseminating actionable information to Joint Forces in a distributed, open-architecture environment. E-2D is also a cornerstone of the Naval Integrated Fire Control – Counter Air system of systems capability.

Utilizing the newly developed AN/APY-9 Mechanical/Electronic Scan Array radar and the Cooperative Engagement Capability system, the E-2D AHE works in concert with tactical aircraft and surface-combatants equipped with the Aegis combat system to detect, track and defeat air and cruise missile threats at extended ranges.

The Fiscal Year 2018 President's Budget requests $292.5 million in RDT&E,N for continuation of added capabilities, to include Aerial Refueling, Secret Internet Protocol Router chat, Advanced Mid-Term Interoperability Improvement Program, Multifunctional Information Distribution System /Joint Tactical Radio System Tactical Targeting Network Technology, Counter Electronic Attack, Sensor Netting, and Data Fusion, Navigation Warfare, Fighter to Fighter Backlink, ALQ217 Electronic Support Measures, and Crypto Modernization/Frequency Remapping. In the fifth year of a 26 aircraft Multi-Year Procurement (MYP) contract covering Fiscal Years 2014-2018, the budget also requests $835.9 million in APN for five FRP Lot 6 aircraft and Advance Procurement for Fiscal Year 2019 FRP Lot 7 aircraft.

ASSAULT SUPPORT AIRCRAFT

MV-22/CMV-22

The Fiscal Year 2018 President's Budget requests $171.4 million in RDT&E,N for continued product improvements, including continued development of a Navy variant, the CMV-22B; and $706.7 million in APN for procurement of 6 Lot 22 CMV-22s.

The DoN begins procurement of the Navy CMV-22B variant in support of the Carrier On-Board Delivery mission in Fiscal Year 2018 which represents the first year of the next V-22 MYP contract (MYP III). The proposed follow-on MYP III contract will span seven years (Fiscal Years 2018-2024) and buy out the remaining domestic aircraft program of record. Fiscal Year 2018 President's Budget requests will fully fund Lot 22 and procure long-lead items for Fiscal Year 2019 Lot 23 CMV-22 aircraft. The request also includes $228.3 million to support Operations and Safety Improvement Programs (OSIPs), including Correction of Deficiencies, Readiness improvements, Common Configuration, and Aerial Refueling.

MV-22 Osprey vertical flight capabilities, coupled with the speed, range, and endurance of fixed-wing transports, continue to enable effective execution of current missions that were previously unachievable. The MV-22 fleet continues executing at a high operational tempo consisting of multiple MEU deployments and two Special Purpose MAGTF - Crisis Response (SPMAGTF-CR) deployments in support of AFRICOM and CENTCOM. During 2016, the 15th of 18 planned active component squadrons met Full Operational Capability (FOC), with the 16th scheduled for FOC in June 2017. This marks the beginning of MV-22 capacity catching up to operational demand requirements. To date, 293 of 360 MV-22s have been delivered and 52 of 53 AFSOC CV-22s have been delivered. The V-22 program focus establishes a third MYP for production aircraft, sustains Fleet aircraft, improves aircraft readiness, reduces operating costs, and expands the domestic and international business base. Both the MV-22 and CV-22 continue to meet all cost, schedule and performance requirements.

MYP III continues affordable procurement, provides stability to industry and maintains a production line and contractual foundation to attract future V-22 international sales/customers. Continuing procurement under a MYP is particularly beneficial to the supplier base. It provides long-term stability and generates lower costs that may provide incentive for international V-22 customers. The program's first Foreign Military Sales program, 17 aircraft with the Government of Japan, was established under MYP II. The final four (of 17 aircraft) are planned to be included in the Fiscal Year 2018 procurement contract.

Due to extremely high CCDR MV-22 demand and operational tempo, the mission capability (MC) aircraft readiness rates have not improved as desired. The primary contributor to lower than planned MC rates is our ability to train and retain enlisted maintainers with the requisite qualifications needed to sustain the high demand. An equally important secondary contributor is related directly to multiple MV-22 configurations. In an attempt to increase overall readiness, the Marine Corps reduced each of the SPMAGTF-CR to a 0.5 VMM squadron footprint. The Marine Corps plans to allow the "remain behind" element necessary time to develop and train personnel for future deployments and improve the overall MV-22 readiness and MC rates.

Marine Aviation commissioned an Osprey Independent Readiness Review which identified a number of factors driving down MV-22 readiness. The major factor identified was the excessive number of aircraft configurations that resulted from years of concurrently incorporating engineering changes and reliability improvements during aircraft production. The Department's "Common Configuration, Readiness and Modernization" plan will streamline the total number of MV-22 configurations from 77 to three, simplify the supply system, reduce the number of technical manuals and improve troubleshooting and maintenance procedures. This will decrease maintenance man-hours, increase aircraft availability and reduce total operating costs by approximately $1.5 billion. The Fiscal Year 2018 OSIP provides a necessary and stable source of crucial modification funding as the program continues to implement these readiness and cost reduction initiatives.

Along with the readiness and support initiatives, the Department is adding new capabilities to the MV-22 that will make it more valuable to the CCDRs such as the development of MV-22 Aerial Refueling System which will enable the MV-22 to deliver fuel to other airborne platforms. This is a critical enabler for both shore and sea-based operations and will extend the operational reach of deployed MAGTFs. Initial capability is planned to deliver by the summer of 2019.

Another transformative capability for the entire aviation force is the continued development and integration of Digital Interoperability (DI). A limited DI objective experiment was conducted utilizing a deployed MEU. The results showed promise and informed continued development of this capability. Initial DI fielded capability will consist of a suite of electronics to allow the embarked troop commander and aircrew to possess unprecedented situational awareness via real-time transmission of full motion video and other data generated by multiple air and ground platforms throughout the battlespace. This DI suite will also be able to collect, in real time, threat data gathered by existing aircraft survivability equipment and accompanying attack platforms, thereby shortening the kill-chain against ground and air based threats.

The MV-22 is the assault support platform of choice for all CCDRs. From MEUs to SPMAGTF-CR, the speed, range, and aerial refueling capability allow the Osprey to remain postured in strategic locations throughout the world, ready and poised to quickly support Marines Corps operations wherever they are required.

CH-53K Heavy Lift Replacement Program

The Fiscal Year 2018 President's Budget requests $341.0 million in RDT&E,N to continue the Engineering Manufacturing Development (EMD) phase of the CH-53K program and $756.4 million in APN for Low Rate Initial Production (LRIP) Aircraft (Lot 2), including Advance Procurement and initial spares.

The CH-53K achieved Milestone C, receiving an Acquisition Decision Memorandum April 3, 2017, authorizing LRIP. To date, four Engineering Development Model aircraft have accumulated over 450 test flight hours, completed the first 'Operational Test Assessment' ahead of schedule and set a U.S. Heavy Lift record with an 89.5K Maximum Gross Weight lift.

During Fiscal Year 2018, the program will continue to execute developmental test flights, complete the relocation of test assets to NAS Patuxent River, and take delivery of System Demonstration Test Article (SDTA) aircraft (production representative aircraft utilized for Operational Test). Three of the four SDTAs will deliver to NAS Patuxent River to supplement the remainder of developmental test. Marine Test and Evaluation Squadron One will take delivery of the balance of aircraft at Marine Corps Air Station (MCAS) New River to execute publication and maintenance demonstrations prior to Operational Test & Evaluation.

The CH-53K will provide land and sea based heavy-lift capabilities not resident in any of today's platforms and contribute directly to the increased agility, lethality, and presence of joint task forces and MAGTFs. The CH-53K can transport 27,000 pounds of external cargo out to a range of 110 nautical miles under the most extreme operational conditions, nearly tripling the CH-53E's lift capability under similar environmental conditions, while fitting into the same shipboard footprint. The CH-53K will provide an unparalleled lift capability under high-altitude and hot weather conditions and greatly expand the CCDRs operational reach and flexibility.

Compared to the CH-53E, maintenance and reliability enhancements of the CH-53K will improve aircraft availability and ensure cost effective operations. Additionally, survivability and force protection enhancements will dramatically increase protection for both aircrew and passengers. Expeditionary heavy-lift capabilities will continue to be critical to successful land and sea-based operations in future anti-access, area-denial environments, enabling sea-basing and the joint operating concepts of force application and focused logistics.

CH/MH-53E

As the CH-53E approaches 30 years of service, the community has accumulated over 95,000 combat flight hours in support of various combat operations. The unprecedented operational demand of this aircraft (peaking at three times the published utilization rate) has degraded the material condition of our heavy lift assault support aircraft sooner than expected. This makes them more challenging to maintain and underscores the importance of its replacement, the CH-53K King Stallion. We have instituted a fleet wide "reset" of the CH-53E inventory to ensure we extract maximum utility and readiness until the transition to the CH-53K occurs.

The MH-53E continues to perform its primary mission of airborne Mine Countermeasures (MCM) as well as transport of cargo and personnel. Over the past 12 years the MH-53E community has accumulated 84,131 flight hours. It too is approaching 30 years of service life and continues to be a challenging asset to maintain. MCM operations put added stress on these airframes. These aircraft are planned to remain in service until they are replaced by the Littoral Combat Ship (LCS) with its MCM mission package systems.

To keep the CH-53E and MH-53E viable through their remaining services lives, the 2018 President's Budget requests $37.0 million in APN and $5.1 million in RDT&E,N. The requested funding provides for critical capabilities, including Condition Based Maintenance software upgrades, finishing Kapton wiring replacement installations, improved engine nacelles, non-recurring engineering cockpit upgrades, Embedded Global Positioning System/Inertial Navigation System, T-64 engine reliability improvements, critical survivability upgrades, satellite communications kits and Phase I of CH-53E's Degraded Visual Environment capability. These critical safety and avionics upgrades will address obsolescence issues within the cockpit and increase overall situational awareness and mission effectiveness.

ATTACK AND UTILITY AIRCRAFT

UH-1Y / AH-1Z

Marine Corps Venom and Viper utility and attack aircraft have been critical to the success of the Marines in harm's way and have flown over 162,000 hours over the past decade. The Fiscal Year 2018 President's Budget requests $61.3 million in RDT&E,N for continued product improvements and $822.4 million in APN for 22 AH-1Z aircraft and system improvements. This budget reflects a deliberate decision to fund readiness through a five aircraft procurement reduction.

As of April 2017, 210 aircraft are operational within the Marine Force (146 UH-1Ys and 64 AH-1Zs). An additional 72 aircraft are on contract and in production, to include the first three of 12 Pakistan Foreign Military Sales aircraft. Lot 1-7 (Fiscal Years 2004-2010) aircraft deliveries are complete for both the UH-1Y and AH-1Z. Lot 8, 9, and 10 (Fiscal Years 2011-2013) deliveries are complete for the UH-1Y. Lot 11 UH-1Y deliveries are in progress and ahead of schedule. Additionally, the Czech Republic signed a Letter of Request for Letter of Acceptance in April 2017 for 12 UH-1Ys, which will be placed on contract in Fiscal Year 2018.

The H-1 Upgrades program is integrating both the UH-1Y and AH-1Z into the DI environment established throughout the MAGTF. With the integration of IT II EW pod, the Marine Corps' Light Attack Helicopter Squadron community will be able to provide MAGTF Commanders with all six functions of Marine Aviation, further increasing capability and flexibility. Additionally, these aircraft will incorporate Software Reprogrammable Payloads (SRP), which enables utilization of diverse networks and waveforms, thereby enabling maneuverability within the EW spectrum. SRP will employ systems such as Link-16, Tactical Targeting Network Technology, Adaptive Networking Wideband Waveform, and the Soldier Radio Waveform.

MH-60 (Overview)

MH-60 Seahawks have consistently met readiness and operational commitments. There will be 38 Navy Seahawk squadrons, with 275 MH-60S and 280 MH-60R aircraft, when transitions from the SH-60B, SH-60F, and HH-60H are complete. The last MH-60S delivered in January of 2016 and MH-60R deliveries are projected to continue into Fiscal Year 2018. The production program continues to deliver on cost and on schedule. Over the last twelve years of combat operations, deployed ashore and aboard our aircraft carriers, amphibious ships, and surface combatants at sea, Navy H-60 helicopters have provided vital over-watch and direct support to troops in combat across multiple theaters of operation and a variety of mission areas; including support for Surface Warfare (SUW), Anti-submarine Warfare (ASW), special operations forces, mine warfare, logistics support and humanitarian assistance/disaster relief.

The MH-60R Multi-Mission Helicopter provides Carrier Strike Group protection and adds significant capability in its primary mission areas of ASW, EW and SUW. The MH-60R is the sole organic air ASW asset in the Carrier Strike Group and serves as a key contributor to theater level ASW. The MH-60R also employs advanced sensors and communications to provide real-time battlespace management with a significant, active or passive, over-the-horizon targeting capability, as well as Fast Attack Craft/Fast In-shore Attack Craft threat response capabilities. Secondary mission areas include Search and Rescue, Vertical Replenishment, Naval Surface Fire Support, Logistics Support, Personnel Transport and Medical Evacuation.

The MH-60S supports Carrier and Expeditionary Strike Groups, Combat Logistics Ships, and LCS Surface Warfare and Mine Countermeasures variants in the mission areas of SUW, Strike Warfare, Combat Search and Rescue, Vertical Replenishment.

The Fiscal Year 2018 President's Budget requests $11.3 million in RDT&E,N across the FYDP for an MH-60S SLAP. SLAP will inform the Department on what will be required to extend the MH-60S airframe service life beyond 2030. The program will initially focus on the air vehicle and include a Fatigue Life Assessment, Dynamic Component, and Subsystem Analysis to inform SLEP requirements.

The Budget request includes $5.4 million in RDT&E,N to support the MH-60 test program and other improvements. The MH-60 test program consists of numerous system upgrades and Pre-Planned Product Improvements, and include the Multifunctional Information Distribution System - Low Volume Terminal Block Upgrade 2, the VHF Omnidirectional Ranging/Instrument Landing System, System Configuration 18 enhancements, MH-60S fixed forward-firing weapon/rocket corrections of deficiencies, and commencement of initial studies for a MH-60 Mid-Life Upgrade. These investments improve MH-60S lethality and provide forward-deployed capabilities to defeat area-denial strategies and allow joint forces to project and sustain power.

EXECUTIVE SUPPORT AIRCRAFT

VH-3D/VH-60N Executive Helicopter Series

The VH-3D and VH-60N are safely performing the Executive Lift mission worldwide. As these aircraft continue to provide seamless vertical lift for the President of the United States, the DoN works closely with HMX-1 and industry to sustain these aircraft until a Presidential Helicopter Replacement platform (VH-92A) is fielded.

The Fiscal Year 2018 President's Budget requests an investment of $38.8 million of APN to continue programs that will ensure the in-service Presidential fleet remains safe and reliable. Ongoing efforts include a Communications Suite Upgrade (Wide Band Line of Sight) that provides persistent access to the strategic communications network, the continuing Structural Enhancement Program necessary to extend the service life, and Obsolescence Management needed to sustain and improve system readiness for both VH-60N and VH-3D platforms. The Cabin Interior and Environmental Control System upgrade is a critical obsolescence management effort for the VH-3D, reducing aircraft operational weight and improving maintainability. Where appropriate, technology updates for legacy platforms will be directly leveraged for the benefit of the VH-92A program.

VH-92A Presidential Helicopter Replacement Aircraft

The Fiscal Year 2018 President's Budget requests $451.9 million in RDT&E,N to continue Engineering Development Model (EDM) activities, to include, contractor test for airworthiness certification and modifications of EDM and System Demonstration Test Article aircraft. The Sikorsky S-92A aircraft will be used to execute the acquisition strategy of integrating mature subsystems into an air vehicle that is currently in production. Significant progress has been made in the past year: completion of the System Critical Design Review in July 2016; continued progress of the test aircraft build with first flight and Contractor Test beginning July 2017; and the projected induction of the first of four S-92A aircraft into the modification process in May to become the SDTA aircraft that will support IOC. Government ground and flight testing is planned to commence in 2018. The first four of the planned operational inventory of 21 aircraft are planned to achieve IOC in 2020.

FIXED-WING AIRCRAFT

KC-130J

The DoN continues to procure two KC-130Js per year, and will continue product improvements. Targeted improvements include aircraft survivability through advanced electronic countermeasure modernization and obsolescence upgrades to the Harvest HAWK ISR/Weapon Mission Kit.

Fielded throughout our active force, the KC-130J brings increased capability, performance and survivability with lower operating and sustainment costs for the MAGTF. Forward deployed in support of ongoing operations since 2005, the KC-130J continues to deliver Marines, fuel and cargo whenever and wherever needed. Today, the KC-130J remains in high demand, providing tactical air-to-air refueling, assault support, CAS and Multi-sensor Imagery Reconnaissance (MIR) capabilities in support of SPMAGTFs and deployed MEUs.

First deployed in 2010, the roll-on/roll-off Harvest HAWK Mission Kit for the KC-130J continues to provide extended MIR and CAS capabilities. With almost 7,000 hours flown, 210 Hellfire missiles, and 91 Griffin missile combat engagements, this expeditionary mission kit has proven its worth and made the KC-130J even more indispensable for Marines on the ground. All six mission kits have been fielded, and the requested funding in the Fiscal Year 2018 budget request will be used to maintain operational relevance of this mission system through compatibility with additional Hellfire variants and an improved full motion video data-link.

The Marine Corps has funded 66 of the 79 KC-130J aircraft through the current FYDP. The 3 aircraft included in the Fiscal Year 2013 budget would have completed the Active Component (AC) requirement of 51 aircraft. However, in 2014 the Marine Corps began using the AC backup aircraft to accelerate the Reserve Component (RC) transition from the legacy KC-130T aircraft to the more capable and efficient KC-130J. The aircraft requested in the Fiscal Year 2018 President's Budget will continue to increase KC-130J inventory as we strive to achieve FOC in the RC. Delays in procurement would force the Marine Corps to sustain the KC-130T aircraft longer than planned at an increased cost and incur additional manpower issues.

It is also important to note that the USAF C-130J procurement is anticipated to end in 2023. If the Marine Corps procure KC-130Js at a rate of two per year, we will have approximately ten aircraft remaining to procure after Fiscal Year 2023 in order to reach the POR of 79 aircraft. The loss of USAF aircraft quantities and the uncertainty of additional Foreign Military Sales may result in a significant unit cost increase for these final aircraft.

MARITIME SUPPORT AIRCRAFT

P-8A Poseidon

The P-8A Poseidon recapitalizes the ASW, Anti-Surface Warfare (ASuW) and armed ISR capabilities from the aging P-3C Orion. The P-8A combines the proven reliability of the commercial 737 airframe with avionics that enable integration of modern sensors and robust military communications. The first P-8A operational deployment was completed in June 2014, with continuous deployments to both 7th Fleet and 6th Fleet underway. As of April 2017, seven of twelve fleet squadrons have completed transition and an eighth is underway. All squadrons are scheduled to complete transition by Fiscal Year 2020. The P-8A program is meeting all cost, schedule and performance parameters in accordance with the approved Acquisition Program Baseline. It has achieved and surpassed reliability standards for operational availability and delivered forward commanders unprecedented capability.

Each of the 54 fleet aircraft delivered early or on time. Lot 6 and Lot 7 are under contract, including eight aircraft for the Royal Australian Air Force, our cooperative partner. Lots 8-10 will include nine aircraft for the United Kingdom and five for the Royal Norwegian Air Force. In Fiscal Year 2018, our request is for $1.386 billion in APN for seven aircraft and $181.7 million in RDT&E,N for aircraft updates to include the addition of Networked Enabled Weapons capabilities.

The first planned upgrade for the P-8A, Increment 2, added a broad-area, multi-static acoustic (MAC) ASW capability to the aircraft. This capability significantly increased the P-8A ASW search rates in harsh, littoral environments. The capability is scheduled to receive regular incremental upgrades over the next five years in order to pace the threat and improve the aircraft's search capability. MAC completed Follow-On Operational Test & Evaluation in April 2015 and has been delivered to the Fleet. Separately, Increment 2 integrates a High Altitude ASW Weapons Capability under a contract awarded in December 2014, in support of a planned 2020 fleet introduction.

P-3C Orion

The aging P-3C fleet will continue to provide critical ASW, ASuW and ISR support for operations worldwide until the fleet completes transition to P-8A. The Fiscal Year 2018 budget request provides $0.7 million to manage P-3C aircraft mission systems obsolescence and $1.4 million to fund the P-3 Fatigue Life Management Program in order to maintain sufficient capacity to complete the transition to P-8A.

EP-3 Aries

The EP-3E Aries is the Navy's only Maritime ISR and Signals Intelligence (SIGINT) platform. The Joint Airborne SIGINT Common Configuration includes Multi-INT sensors, robust communication, and data links employed by the venerable P-3 air vehicle to ensure effective fleet support across the full spectrum of military operations. The Fiscal Year 2011 National Defense Authorization Act directed the Navy to sustain EP-3E airframe and associated mission systems to minimize SIGINT capability gaps until the systems are fully recapitalized with a system or family of systems that in aggregate provide equal or better capability and capacity. The Navy's family of systems approach to ISR shifts the focus from platforms to payloads to deliver increased capacity and persistence by the end of this decade. The EP-3 Fiscal Year 2018 budget request of $14.5M (Baseline and OCO) reduces risk compared to previous fiscal years while the Navy continues to collaborate with the Joint Staff and DoD to optimize the future of ISR.

UNMANNED AIRCRAFT SYSTEMS (UAS)

The DoN has placed a priority on the development of unmanned systems leading to a fully integrated manned and unmanned fleet. Unmanned technology will not replace our Sailors and Marines; instead it will unlock their full potential as we integrate this technology within our total force.

MQ-4C Triton UAS

The Fiscal Year 2018 President's Budget requests $84.1 million in RDT&E,N to continue Triton baseline development activities; $229.4 million in RDT&E,N for Triton modernization; and $676.3 million of APN for procurement of the third lot of LRIP aircraft and spares, retrofit of the LRIP Lot 1 aircraft to the Multi-INT configuration, and for procurement of long lead materials for the fourth lot of LRIP aircraft.

The MQ-4C Triton is a key component of the Navy Maritime Patrol Reconnaissance Force. Its persistent sensor dwell, combined with networked sensors, will enable it to effectively meet ISR requirements in support of the Navy Maritime Strategy. Triton will start establishing five globally-distributed, persistent Maritime ISR orbits beginning in Fiscal Year 2018, as part of the Navy's Maritime ISR&T Transition Plan. MQ-4C Triton test vehicles have completed over 110 test flights as of April, 2017, and will complete sensor and performance flight testing this fall in support of establishing an early operational capability in the Pacific next year. Milestone C was successfully completed in September 2017, and the program has entered the production and deployment phase.

The Navy currently maintains an inventory of four RQ-4A Global Hawk Block 10 UAS, as part of the BAMS Demonstrator, or BAMS-D program. These aircraft have been deployed to CENTCOM's AOR for over eight years. BAMS-D recently achieved over 23,000 flight hours in support of CENTCOM ISR tasking.

MQ-25 Stingray

MQ-25 will deliver the Navy's first carrier-based unmanned aircraft to function primarily as a mission tanker to extend the range and reach of the CVW with secondary recovery tanking and ISR capabilities. MQ-25 will reduce current use of F/A-18E/Fs as CVW tankers and extend F/A-18E/F service life. As a secondary mission, MQ-25 will provide the Carrier Strike Group Commander an organic, persistent ISR capability for maritime domain awareness. The Fiscal Year 2018 President's Budget requests $222.2 million in RDT&E,N for MQ-25 developmental activities and the Air System Engineering and Manufacturing Development contract award.

MQ-8 Fire Scout

The MQ-8 Fire Scout is a rotary-wing system that includes two airframe types, the MQ-8B and MQ-8C. The MQ-8C is a larger, more capable and more cost-effective airframe that uses the same mission control system, avionics and payloads as the MQ-8B. The system is designed to operate from any suitably-equipped air-capable ship, carry modular mission payloads, and operate using the Tactical Control System and Line-Of-

Sight Tactical Common Data Link. The Fiscal Year 2018 President's Budget requests $62.7 million of RDT&E,N to continue hardware and software modifications, payload integration, cyber vulnerability closure and safety capability improvements such as a backup landing system and collision avoidance system. The request for $85.4 million in APN procures four MQ-8 mission control systems, MQ-8C AESA Radar kits, ancillary shipboard equipment, trainers and aircraft support equipment, technical support, modifications based on engineering changes, and logistics products to outfit suitably-equipped air-capable ships and train the associated Aviation Detachments.

The MQ-8B has completed 11 operational deployments and flown more than 16,000 operational hours, including deployments to Afghanistan, deployments on Navy Frigates, and deployments aboard LCS supporting Special Operations Forces and Navy operations. The MQ-8B is currently deployed on USS CORONADO (LCS-4) with HSC-23 in a composite aviation detachment with an MH-60S. This detachment represents the first deployment of an MQ-8B with a maritime search radar capability. HSC-21, located in San Diego, California, is currently working up for a Fiscal Year 2018 employment onboard USS INDEPENDENCE (LCS-2) marking the first deployment of the Coastal Battlefield Reconnaissance and Analysis MCM payload. HSC-22, located in Norfolk Virginia, has been identified as the MQ-8 introductory squadron for the east coast and will deploy onboard the USS DETROIT (LCS-7) in early 2018.

The MQ-8C Fire Scout has flown more than 800 flight hours conducting developmental and land-based operational testing including dynamic interface testing on LCS-8 in April 2017. The program begins Initial Operational Test & Evaluation in the first quarter of Fiscal Year 2018. The Navy is continuing efforts to integrate an AESA radar capability into the MQ-8C and is planning to integrate the APKWS II weapon system and future MCM payloads. The Fire Scout program will continue to support integration and testing for LCS-based Surface Warfare and MCM mission modules.

Tactical Control System (TCS)

The Fiscal Year 2018 President's Budget requests $7.8 million in RDT&E,N for the MQ-8 System's Tactical Control System (TCS). TCS provides a standards-compliant open architecture with scalable command and control capabilities for the MQ-8 Fire Scout system. In Fiscal Year 2018, TCS will continue transition of the Linux operating system to a technology refreshed mission control system, and enhance the MQ-8 System's Automatic Identification System and sensor track generation integration with ship systems. The Linux operating system conversion overcomes hardware obsolescence issues with the Solaris based control stations and provides lower cost software updates using DoD common application software. In addition, the TCS Linux upgrade will enhance collaboration with the Navy's future UAS Common Control System.

RQ-21A Blackjack

The Fiscal Year 2018 President's Budget requests $13.7 million in RDT&E ($4.8 million USN, $8.9 million USMC); $4.8 million in APN for support of Naval Special Warfare; and $86.2 million in PMC for four expeditionary RQ-21A systems (which includes 20 air vehicles) to address Marine Corps ISR capability requirements. This Group 3 UAS provides persistent ship and land based ISR support for expeditionary tactical-level maneuver decisions and unit level force defense and force protection missions. Blackjack entered LRIP in 2013, completed Initial Operational Test & Evaluation in the second quarter of Fiscal Year 2015, and reached IOC in January 2016. FRP was approved in the fourth quarter of Fiscal Year 2016.

The RQ-21 completed its first combat deployments in 2016 with support to the 24th and 22nd MEU and Marine Corps Special Operations Command operations in Operation INHERENT RESOLVE. The Blackjack has flown over 700 sorties and 3940 hours in support of the MAGTF.

The RQ-21's current configuration includes full motion video, communications relay package and automatic identification systems. The air vehicle's bay allows for rapid deployment of signals intelligence payloads. The Marine Corps is actively pursuing technological developments for the RQ-21A system in an effort to provide the MAGTF and Marine Corps Special Operations Command with significantly improved capabilities. Initiatives include over-the-horizon communication and data relay ability to integrate the system into future networked digital environments; electronic warfare and cyber payloads to increase non-kinetic capabilities; and change detection radar and moving target indicators to assist warfighters in battlespace awareness and force application.

MAGTF Expeditionary UAS (MUX)

As the Marine Corps recapitalizes toward a more diverse, lethal, amphibious and middleweight expeditionary force, the Marine Corps will require a UAS that is network-enabled, digitally interoperable, and built to execute responsive, persistent, lethal, and adaptive full-spectrum operations. A MUX is planned to be the system that will provide the MEF/MEB-sized MAGTF with an advanced multi-mission platform.

The Fiscal Year 2018 budget requests $5.0 million in RDT&E for the MUX program to conduct an AoA and begin development of an acquisition strategy; $3.0 million in RDT&E for KMAX operations in support of MUX technology demonstrations and Concept of Operation development (included under the MUX line).

The MUX Initial Capabilities Document was approved by the Joint Requirements Oversight Council on October 4, 2016. The AoA study plan and guidance are being developed with OSD(CAPE). The AoA is projected to be completed by the fourth quarter of Fiscal Year 2018.

MUX supports the Marine Corps Operating Concept by significantly mitigating or eliminating the following MAGTF gaps: EW, ISR, Command, Control and Communications (C3) DI, Aerial Escort, all weather, persistent CAS and Deep Air Support, Airborne Early Warning, and Tactical Cargo Distribution. MUX will be a long range (690+ NM), persistent (24+ hours) UAS capable of complimenting MV-22 operations and operating from both sea and expeditionary bases.

Common Control System (CCS)

The Fiscal Year 2018 President's Budget requests $39.7 million in RDT&E,N for the Common Control System (CCS). The primary mission of CCS is to provide common control across the Navy's unmanned systems (UxS) portfolio to add scalable and adaptable warfighting capability, implement robust cybersecurity attributes, leverage existing government owned products, eliminate redundant software development efforts, consolidate product support, encourage innovation, improve cost control, and enable rapid integration of UxS capabilities across all domains: Air, Surface, Sub-Surface, and Ground. CCS leverages existing Government owned software to provide UxS Vehicle Management (VM), Mission Management (MM) and Mission Planning (MP) capabilities. CCS uses an open and modular business model and is being developed initially as Government Furnished Information/Equipment for the MQ-25 and for follow-on use with Triton and Fire Scout. In Fiscal Year 2018, CCS Increment I will continue to perform software design, development, integration and test for VM. Concurrently, CCS Increment II will conduct MM/MP requirements development and software design.

SAFETY

Responses to Congressional requests for updates on Naval Aviation safety can be found at Addendum C.

STRIKE WEAPONS PROGRAMS

Cruise Missile Strategy

The DoN has aligned its Cruise Missile Strategy along warfighter domains to pursue maximized lethality while minimizing overall costs to the taxpayer and Department.

The first tenet of our plan is to sustain the Tomahawk cruise missile inventory through its anticipated service-life via a mid-life recertification program (first quarter of

Fiscal Year 2019 start). This recertification program will increase missile service-life by an additional 15-years (total of 30-years) and enable the Department to support Tomahawk in our active inventory through the mid-late 2040s. In concert with our recertification program we will integrate modernization and technological upgrades and address existing obsolescence issues. In addition, we are developing a Maritime Strike Tomahawk capability to deliver a long-range anti-surface warfare capability.

Second, we will field the Long Range Anti-Ship Missile (LRASM) as the air-launched Offensive Anti-Surface Warfare (OASuW)/Increment 1 material solution to meet near to mid-term anti-surface warfare threats. LRASM is pioneering accelerated acquisition processes in accordance with DoD-5000.02 (Model 4). Currently, we anticipate LRASM to meet all Joint Chiefs of Staff approved warfighting requirements, deliver on-time, and cost within approximately one-percent of its original program cost estimate.

We also plan to develop follow-on next generation strike capabilities. We intend to develop an air-launched OASuW/Increment 2 weapon to address long-term ASuW threats and a surface and submarine launched Next Generation Land Attack Weapon (NGLAW). NGLAW will have both a long-range land strike and maritime ASuW capability that initially complements, and then replaces, the highly successful Tomahawk Weapon System.

To the maximum extent possible, the DoN plans to utilize common components and component technologies (e.g. navigation, communications, seeker, guidance and control) to reduce cost, shorten development timelines, and promote interoperability. Based on performance requirements and launch parameters, next generation strike capability missile airframes and propulsion systems will differ between the air-launched and sea-launched weapons.

Tactical Tomahawk (TACTOM) BLK IV Cruise Missile

The Fiscal Year 2018 President's Budget requests $234.5 million in WPN for procurement of an additional 100 TACTOM weapons and associated support to include replacement of weapons launched in combat (Syria), $31.7 million in OPN for the Tomahawk support equipment, and $114.8 million in RDT&E,N for capability updates of the weapon system. WPN resources will be for the continued procurement of this versatile, combat-proven, deep-strike weapon system in order to meet ship load-outs and combat requirements. OPN resources will address the resolution of Tactical Tomahawk Weapons Control System obsolescence, Tomahawk Theater Mission Planning Center (TMPC) complexity and usability issues, interoperability, and information assurance mandates. RDT&E,N resources will be used to develop navigation system improvements and communications upgrades to improve TACTOMs performance in Anti-Access/Area Denial environments, as well as development of a seeker to enable TACTOM to engage maritime targets, and the development and integration of a multiple effects warhead.

Tomahawk provides an attack capability against fixed and mobile targets, and can be launched from both Ships and Submarines. The current variant, TACTOM, preserves Tomahawk's long-range precision-strike capability while significantly increasing responsiveness and flexibility. TACTOM's improvements include in-flight retargeting, the ability to loiter over the battlefield, in-flight missile health and status monitoring, and. Other Tomahawk improvements include rapid mission planning and execution via Global Positioning System (GPS) onboard the launch platform and improved anti-jam GPS.

Tomahawk Theater Mission Planning Center (TMPC)

The Fiscal Year 2018 President's Budget for TMPC requests $18.8 million in RDT&E,N and $41.5 million in OPN. TMPC is the mission planning and strike execution segment of the Tomahawk Weapon System. TMPC develops and distributes strike missions for the Tomahawk Missile; provides for precision targeting, weaponeering, mission and strike planning, execution, coordination, control and reporting. TMPC provides CCDRs and Maritime Component Commanders the capability to plan and/or modify conventional Tomahawk Land-Attack Missile missions. TMPC is a Mission Assurance Category 1 system, vital to operational readiness and mission effectiveness of deployed and contingency forces. RDT&E,N efforts will address National imagery format changes, update Tomahawk navigation and accuracy algorithms - to include operations in the maritime and/or Anti-Access Area Denial environments, upgrade obsolete Tomahawk Cruise Missile Communications and initiate a Tomahawk seeker integration into the TMPC mission planning environment. OPN resources will enable the Navy to continue software engineering efforts associated with Tomahawk Missile Modernization, upgrade unsupportable and obsolete TMPC software to ensure compliance with DoD cybersecurity mandates, and implement the TMPC Enterprise Network to allow for rapid delivery of security policies, cybersecurity software patches and anti-virus definitions. All of these upgrades are critical for the support of over 180 TMPC operational sites worldwide, afloat and ashore, to include: Cruise Missile Support Activities (inclusive of STRATCOM), Tomahawk Strike and Mission Planning Cells (5th, 6th, 7th Fleet), Carrier Strike Groups, Surface and Subsurface Firing Units and Labs/Training Classrooms.

Offensive Anti-Surface Warfare (OASuW) Increment 1 (Long Range Anti-Ship Missile (LRASM))

OASuW/Increment 1 (LRASM) will provide CCDRs the ability to conduct ASuW operations against high-value surface combatants protected by Integrated Air Defense Systems with long-range Surface-to-Air-Missiles and deny adversaries the sanctuary of maneuver against 2018-2020 threats. The program is scheduled to achieve Early Operational Capability on the Air Force B-1 by the end of Fiscal Year 2018 and Navy F/A-18E/F by the end of Fiscal Year 2019.

The Fiscal Year 2018 President's Budget request contains $160.7 million in RDT&E,N for LRASM development and testing and $74.7 million in WPN to purchase 25 LRASM All-Up-Round weapons. OASuW Increment 1 (LRASM) leverages the Defense Advanced Research Projects Agency weapon demonstration effort.

Offensive Anti-Surface Warfare (OASuW) Increment 2

OASuW/Increment 2 is required to deliver the long-term air-launched ASuW capability to counter 2024 (and beyond) threats. The Department continues to plan for OASuW/Increment 2 to be determined via full and open competition. Full OASuW/Inc. 2 capability is delayed until at least Fiscal Year 2026 (est.).

Next Generation Land Attack Weapon (NGLAW)

The Next Generation Land Attack Weapon (NGLAW) will provide the next generation of long-range, kinetic strike to destroy high-priority fixed, stationary and moving targets – as well as those targets hardened, defended or positioned at ranges such that engagement by aviation assets would incur unacceptable risk. NGLAW will be capable of kinetic land and maritime attack from surface and sub-surface platforms and initially complement, and then eventually replace, the Tomahawk Weapon System. IOC is planned for the 2028-2030 timeframe (est.).

On November 28, 2016, the Under Secretary of Defense approved Navy's entry into the MS-A phase and authorized initiation of an AoA. Fiscal Year 2018 resources totaling $9.9 million begins the transition from the analysis phase to planning for a formal program of record.

Sidewinder Air-Intercept Missile (AIM-9X)

The Fiscal Year 2018 President's Budget requests $ 42.9 million in RDT&E,N and $79.7 million in WPN for this joint DoN and USAF program. RDT&E,N will be applied toward the Engineering Manufacturing Development phase of critical hardware obsolescence redesign and Developmental Testing of Version 9.4 missile software, both part of the AIM-9X/Block II System Improvement Program (SIP) III. Navy also continues the design and development of Insensitive Munitions improvements in accordance with direction from the Joint Chiefs of Staff. WPN funding is requested to procure a combined 185 All-Up-Rounds and Captive Air Training Missiles and associated missile-related hardware. The AIM-9X Block II/ II+ Sidewinder is the newest in the Sidewinder family and is the only short-range infrared air-to-air missile integrated on Navy, Marine Corps, and USAF strike-fighter aircraft. This fifth-generation weapon incorporates high off-boresight acquisition capability and increased seeker sensitivity through an imaging infrared focal plane array seeker with advanced guidance processing for improved target acquisition; data link capability; and advanced thrust vectoring technology to achieve superior maneuverability and increase the probability of intercept of adversary aircraft.

Advanced Medium-Range Air-to-Air Missile (AMRAAM/AIM-120D)

The Fiscal Year 2018 President's Budget requests $25.4 million in RDT&E,N for continued software capability enhancements and $197.1 million in WPN for 120 All-Up-Rounds and associated missile-related hardware. AMRAAM is a joint USAF and DoN weapon that counters existing aircraft and cruise-missile threats. It uses advanced counter-electronic attack capabilities at both high and low altitudes, and can engage targets from both beyond visual range and within visual range. AMRAAM provides an air-to-air first look, first shot, first kill capability, while working within a networked environment in support of the Navy's Theater Air and Missile Defense Mission Area. RDT&E,N will be applied toward critical hardware obsolescence through the Form, Fit, Function, Refresh (F3R) redesign effort as well as software upgrades to counter emerging Electronic Attack threats for AIM-120C/D missiles. Production challenges linked to the F3R program forced the Navy to reduce its planned procurement of AMRAAM in Fiscal Year 2018.

Small Diameter Bomb II (SDB II)

The Fiscal Year 2018 President's Budget requests $112.8 million in RDT&E,N for continued development of the USAF-led Joint Service SDB II weapon and Joint Miniature Munitions Bomb Rack Unit (JMM BRU) programs and $21.0 million in WPN to procure 90 All-Up-Round weapons. Using multi-mode seeker and two-way data-link capabilities, SDB II provides an adverse weather, day or night standoff capability against mobile, moving, and fixed targets, and enables target prosecution while minimizing collateral damage. SDB II will be integrated into the internal carriage of both DoN variants of the Joint Strike Fighter (F-35B/F-35C) and externally on the Navy's F/A-18E/F via the JMM BRU (BRU-77A). JMM BRU completed Milestone B and entered Engineering Manufacturing Development in August 2015. Both SDB II and JMMU BRU will use an Universal Armament Interface architecture to enable more efficient and less costly future weapon/platform integration.

Advanced Anti-Radiation Guided Missile (AARGM) & AARGM Extended Range

The Fiscal Year 2018 President's Budget requests $6.4 million of RDT&E,N for High-Speed Anti-Radiation Missile (HARM) and AARGM Foreign Material Assessment; $15.2M for AARGM to implement M Code, transition receiver upgrade from ONR efforts and Block 1 follow-on development; $66.3 million of RDT&E,N for AARGM Extended Range (AARGM-ER) development; and $183.4 million of WPN for production of AARGM modification kits for 251 All-Up-Rounds and Captive Training Missiles. The AARGM cooperative program with the Italian Air Force transforms the HARM into an affordable, lethal, and flexible time-sensitive strike weapon system for conducting Destruction of Enemy Air Defense missions. AARGM adds multi-spectral targeting capability and targeting geospecificity to its supersonic fly-out to destroy sophisticated enemy air defenses and expands upon the HARM target set. The program achieved IOC on the F/A-18C/D aircraft in July 2012, with forward deployment to PACOM; integration is complete for AARGM with release of H-8 System Configuration Set for F/A-18E/F and EA-18G aircraft. The AARGM Block 1 software only update will achieve IOC the third quarter of FY 2017. The AARGM-ER modification program, involving hardware and software improvements, began in Fiscal Year 2016. This effort will increase the weapon system's survivability against complex and emerging threat systems and affords greater stand-off range for the launch platform. AARGM-ER will be designed to fit internally in both the F-35A and F-35C, thereby increasing the capability and lethality of the Lightening II weapon system.

Joint Air-to-Ground Missile (JAGM)

The Fiscal Year 2018 President's Budget requests $15.5 million in RDT&E,N to continue a five year integration effort of JAGM Increment 1 onto the Marine Corps AH-1Z and $3.8 million in WPN to support the Fiscal Year 2017 procurement of 96 All-Up-Rounds in order to meet the IOC in Fiscal Year 2020. The Fiscal Year 2017 and Fiscal Year 2018 funding will be used to procure the JAGM LRIP All Up Rounds, Other Production Support, training missiles, production related engineering and logistics to support the procurement in order to meet the IOC.

JAGM is an Army-led, Joint ACAT-1D Major Defense Acquisition Program. JAGM is a direct attack/CAS missile program that will utilize advanced seeker technology to provide fire-and-forget, simultaneous target engagement against land and maritime targets. JAGM will replace the HELLFIRE and TOW II missile systems for the DoN. In November 2012, the Joint Chiefs of Staff authorized the JAGM incremental requirements and revalidated the DoN's AH-1Z Cobra aircraft as a threshold platform. JAGM Increment 1 achieved Milestone B approval in Fiscal Year 2015, a Milestone C (LRIP) is planned for the Fiscal Year 2018 and AH-1Z Cobra/JAGM IOC is planned for Fiscal Year 2020.

Advanced Precision Kill Weapon System II (APKWS II)

The Fiscal Year 2018 President's Budget requests $39.5 million in PANMC for procurement of 1,210 APKWS II Precision Guidance Kits. APKWS II provides an unprecedented precision guidance capability to DoN unguided rocket inventories, improving accuracy and minimizing collateral damage. Program production continues on schedule, meeting the needs of our warfighters in today's theaters of operations. Marine Corps AH-1W and UH-1Y achieved IOC in March 2012 and the Marine Corps AH-1Z platform was certified to fire APKWS II in June 2015. To date, these platforms have expended more than 190 APKWS II weapons during combat missions. The Navy successfully integrated APKWS II on the MH-60S for an Early Operational Capability in March 2014 and fielded a similar effort on the MH-60R in March 2015. A variant of

APKWS II has been integrated onto the AV-8B, A-10 and F-16 aircraft, and is currently being employed in support of Operation INHERENT RESOLVE.

Direct Attack Weapons and General Purpose Bombs

The Fiscal Year 2018 President's Budget requests $108.9 million in PANMC for Direct Attack Weapons and General Purpose bombs and an additional $164.3M specifically to procure 7,209 Joint Direct Attack Munition (JDAM) kits to enhance readiness. In thirty months of Operation INHERENT RESOLVE, DoN aircraft have expended more than three times the number of 500lb JDAM kits than we have procured during the same period. This significant warfighter demand has forced the Navy to reduce the number of 500-pound JDAM available for training in order to preserve warfighting inventory. The OCO request for Fiscal Year 2018 replaces the ordnance expended in the first six months of 2016. While OCO replenishment is helpful, it does not overcome the remainder of the year's expenditures which will continue to exacerbate the current inventory shortfall. Fully funding the General Purpose Bomb line item is critical to sustaining the DoN's inventory for ongoing combat operations and replenishing it for future contingencies.

CONCLUSION

The Department of the Navy continues to instill affordability, strive for stability, and maintain capacity to advance capabilities and meet mission requirements. We remain an agile strike and amphibious power projection force in readiness, and such agility requires that the aviation arm of our naval strike and expeditionary forces remain strong. Mr. Chairman, and distinguished committee members, we request your continued support for the Department's Fiscal Year 2018 budget request for our Naval Aviation programs.

Addendum A
PHYSIOLOGICAL EPISODES

Physiological Episodes (PEs) occur when aircrew experience a decrement in performance, related to disturbances in tissue oxygenation, depressurization or other factors present in the flight environment. PEs are categorized into two general groups, those related to Onboard Oxygen Generation Systems (OBOGS) or pilot breathing gas, and those caused by problems in the Environmental Control System s (ECS), i.e. – unscheduled pressure changes in the flight station. These phenomena jeopardize safe flight.

As a result of physiological episodes, the F/A-18 Program Office (PMA-265) established a Physiological Episode Team (PET) in 2010. In March of 2017, the PET was reorganized to form the PMA-265 Physiological Episode (PE) Integrated Product Team (IPT) to perform a formal Root Cause and Corrective Action analysis of F/A-18A-F and EA-18G events. The F/A-18 PE IPT is a formal partnership between PMA-265 and Boeing, and includes participation from Northrop Grumman, the NAVAIR Engineering Fleet Support Team (FST), NAVAIR 4.3's Environmental Control System (ECS) Team, NAVAIR 4.6's Human Systems Team, and the NAE's Aeromedical Crisis Action Team. The F/A-18 PE IPT works closely with other program offices, cross-service affiliates and industry partners in evaluating each episode for root cause and appropriate corrective action.

The PMA-265 PE IPT is currently addressing hypoxia and decompression events as the two most likely causes of recent physiological episodes in aviators. As symptoms related to depressurization, tissue hypoxia and contaminant intoxication overlap, discerning a root cause is a complex process. Episodes of decompression sickness typically accompany a noticeable loss or rapid fluctuation of cabin pressure, while the cause of hypoxic related events is often not readily apparent during flight or post flight. Reconstruction of the flight event is difficult with potential causal factors not always readily apparent during post-flight debrief and examination of aircraft and aircrew.

Historical data of F/A-18 physiological events prior to May 2010 is based on safety reports. The rate per 100,000 flight hours during FY 2006-FY 2010:

Date Range	F/A-18A-D	F/A-18E-F	EA-18G
FY06	3.66	2.18	0.00
FY07	1.63	3.73	0.00
FY08	3.72	4.28	0.00
FY09	6.19	8.33	0.00
FY10	4.95	11.96	0.00

In May 2010, the Commander, Naval Air Forces directed specific reporting procedures to collect more data on the occurrence of PEs. Following implementation of the new reporting protocol, the rate per 100,000 flight hours beginning in May 2010:

Date Range	F/A-18A-D	F/A-18E-F	EA-18G
05/1/2010 - 10/31/2010	12.20	8.98	0.00
11/1/2010 - 10/31/2011	10.90	8.65	5.52
11/1/2011 - 10/31/2012	16.39	23.35	5.42
11/1/2012 - 10/31/2013	21.01	26.23	9.80
11/1/2013 - 10/31/2014	29.54	26.39	15.05
11/1/2014 - 10/31/2015	30.20	28.02	42.89
11/1/2015 - 10/31/2016	57.24	31.05	90.83

The process for investigating a physiological episode begins with the submission of data describing the event. Engineers from the ECS FST and the Aircrew Oxygen Systems In-Service Support Center work with the squadron maintenance department to identify which components of the aircraft should be removed and submitted for engineering investigation. The squadron flight surgeon also submits data on the medical condition of the pilot and in-flight symptoms that were experienced.

After completion of the component investigations, the incident is examined holistically by members of the engineering teams and Aeromedical specialists to identify the most likely cause of the incident. Of 382 cases adjudicated by the PET so far, 130 have involved some form of possible contamination, 114 involved an ECS component failure, 91 involved human factors, 50 involved an OBOGS component failure, 13 involved a breathing gas delivery component failure, and 76 were inconclusive or involved another aircraft system failure. Of note, some of the events resulted in assignment to more than one category.

T-45 Physiological Episodes

Data recorded since introduction of the T-45 Physiological Event Reporting Protocol form in November 2011 is presented below by calendar year. Prior years' data for T-45 aircraft is incomplete and is not included.

Calendar Year	Calendar year rate per 100K flight hours	Cumulative rate per 100K flight hours
2012	11.86	11.86
2013	16.22	13.94
2014	18.43	15.36
2015	44.99	22.70
2016	46.97	28.01

The process for investigating a physiological episode mimics that being used by the F/A-18 and is also managed by PET. After completion of the component investigations, the incident is examined holistically by members of the PET's engineering teams and aviation medical specialists to identify the most likely cause of the incident. More than one causal factor can be attributed to a single physiological episode event. Of the 79 physiological episode reports adjudicated to date, 24 were assessed to be possible contamination, 12 involved human factors (these may also include incidents of airsickness and vertigo), 12 involved OBOGS component failure, 11 involved a breathing gas delivery failure, three involved cabin integrity, and the remaining 23 were inconclusive or involved another system failure.

Efforts to Mitigate Physiological Episodes on F/A-18 and EA-18G

A variety of actions have been undertaken to address the occurrence of physiological episodes in the F/A-18 / E/A-18G:

1. New maintenance rules for handling the occurrence of specific ECS built-in test faults have been implemented throughout the fleet requiring that the cause of the fault be identified and corrected prior to next flight.
2. Transportable Recompression Systems have been put on forward deployed aircraft carriers to immediately treat aircrew in the event they experience decompression sickness symptoms.
3. Mandatory cabin pressurization testing is now performed on all F/A-18A-F and EA-18G aircraft every 400 flight hours and ECS pressure port testing is performed on all F/A-18A-D aircraft every 400 flight hours. Overhaul procedures for ECS components and aircraft servicing procedures have been improved.
4. Emergency procedures have been revised, all pilots now receive annual hypoxia awareness training, and biennial dynamic training using a Reduced Oxygen Breathing Device to experience and recognize hypoxia symptoms while operating an aircraft simulation.
5. Aircrews are provided portable hypobaric recording watches to alert them when cabin altitude reaches a preset threshold.
6. Internal components of the F/A-18 OBOGS have been redesigned to incorporate a catalyst to prevent carbon monoxide from reaching the pilot and provide an improved capability sieve material (filter). These new OBOGS components have been installed in 84 percent of the in service F/A-18 fleet so far.

7. Improvements to existing maintenance troubleshooting procedures and acceptance and test procedures for reworked components have been incorporated and additional improvements are under evaluation.
8. Hardware and software changes are in work for Super Hornets and Growlers to mitigate cabin pressurization issues due to moisture freezing in the ECS lines.
9. Component redesign, improved performance testing, and newly established life limits will improve component reliability across all F/A-18 configurations.
10. An increased capacity for the emergency oxygen bottles is under contract.
11. Trial sampling efforts for contamination have been conducted at EA-18G squadrons located at NAS Whidbey Island to improve real-time data collection for OBOGS related systems. "Sorbent tubes" which help collect and identify unknown contaminants have been attached to aircrew regulators to collect samples of breathing gas for post-flight analysis of potentially harmful compounds.
12. An ECS laboratory is under construction to improve root cause and correct actions of ECS engineering investigations of fleet events. The projected operational date of the ECS lab is September of 2017.
13. Aircraft are flown with "slam sticks" to track and collect cabin pressure changes over time for rigorous data analysis and to compare data to what the aircrew experienced.
14. Future projects include systematic evaluations of technologies to monitor and detect physiological symptoms.

Efforts to Mitigate Physiological Episodes on T-45

A variety of actions have been undertaken to address the occurrence of physiological episodes in the T-45:

1. Instituted recurring immersion training at all Chief of Naval Air Training sites using Reduced-Oxygen Breathing Devices.
2. Flight manual procedures were updated to optimize crew posture for PE recognition, response, and avoidance.
3. Revised maintenance publications at both the operational and intermediate maintenance levels to increase the minimum oxygen generating performance of the concentrator.

4. Conducted engine wash water intrusion tests to determine if water was entering the OBOGS bleed air. Tests indicated that no water was ingested in the OBOGS bleed air lines.
5. Installed sorbent tubes and hydrocarbon detectors on aircrew to monitor breathing gasses coming off OBOGS. The sorbent tube and HCD are attached to the aircrew vest and ported off the oxygen mask hose.
6. Installed new sieve beds in the Gas Generating Unit (GGU)-7 Oxygen Concentrator. The new sieve beds addressed the possibility of built up contaminants in the sieve bed material by installing all new material, and incorporated a carbon monoxide catalyst to protect against carbon monoxide.
7. Began fielding of new design CRU-123 oxygen monitoring units. A fielded demo unit has over 100 flight hours; up to 15 additional new monitors are expected by the end of May. Thirty additional units will be installed every month thereafter. The new oxygen monitor provides new aircrew alerting if delivery pressure falls, and it records system performance and faults.
8. Initiated requirements analysis for a new OBOGS oxygen concentrator unit.
9. Formed a combined team with Government, Boeing (T-45 OEM), and Cobham (Oxygen Concentrator OEM) members to cooperate on multiple lines of effort to address Physiological Episodes.
10. Conducted multiple rounds of high intensity stress testing of the GGU-7 Oxygen Concentrator at both NAVAIR and Cobham Laboratories to determine concentrator performance outside of the normal operating limits (high temperature and high humidity).
11. NAVAIR released an end to end cleaning procedure for the OBOGS bleed system. Updated regular maintenance procedures to sustain system hygiene. Additional thorough cleaning procedures are being developed.
12. Evaluated the thermal performance of the OBOGS bleed air system by conducting tests on in-service heat exchangers and temperature switches that provide alerts when over-temperature conditions occur.
13. Conducted laboratory testing and on-aircraft fit checks of a new water separator that would be installed in the OBOGS bleed line prior to the OBOGS concentrator to help

guard against water intrusion in the concentrator. This program is currently in the early stages of detailed engineering design.

14. Enhanced data management and collection through initiation of a new data management plan; contracted data analysis support to
15. Developed new test procedures and conducted OBOGS and ECS bleed air contaminant testing on fleet aircraft to establish measurement thresholds and foment a predictive system performance methodology; developed new test sets to assess oxygen system degraded performance.
16. Updated flight and maintenance publications to help prevent inadvertent system damage, ensure leak free system integrity, add periodic inspections, and ensure system cleanliness.

The Department of the Navy remains focused on solving this issue. Fleet awareness is high, protocols are in place and we are focused on mitigating risk, correcting known deficiencies and attacking this issue. Moving forward we will continue to fly while applying every resource to solve this challenging problem.

End of Addendum A

Addendum B
ELECTRONIC WARFARE SUPPLEMENTAL

AN/ALQ-214 - Navy completed testing the upgraded version of the ALQ-214 v4 Integrated Defensive Electronic Countermeasure (IDECM) last year and continues developing software improvements under the Software Improvement Program (SWIP). IOC of SWIP is expected in the second quarter of Fiscal Year 2018. IDECM hardware is currently being installed into deploying F/A-18 E/F aircraft on the planned procurement ramp.

Next Generation Jammer (NGJ) - The first increment of NGJ, which covers a mid-band frequency range, completed its critical design review in May and is on timeline for a Fiscal Year 2021 IOC. OSD established this program as a Skunk Works charter in Fiscal Year 2015 which has allowed a small team of experts to streamline the acquisition process. The Next Generation Jammer Low Band (increment 2) is the next material solution to replace the 40 year old ALQ-99 low band transmitter systems. The acquisition strategy for Low Band (Inc. 2) will be a full and open competition supporting program entry at Milestone (MS) B. Prior to the EMD competition, there will be up to three Demonstration of Existing Technology (DET) contracts awarded as an extension of the Low Band (Inc. 2) program's market research effort. In the execution of the DET contracts, contractors will demonstrate their existing, mature technologies in a relevant environment (i.e. not a technology maturation effort, but rather substantiation of the assertion the technologies of appropriate level of maturity currently exist to support program entry at MS B). Not being awarded a DET contract will not preclude any contractor from submitting a proposal and competing for award of the Low Band (Inc. 2) EMD contract, as, again, it will be a full and open competition. IOC for NGJ Low band is being planned for Fiscal Year 2025.

ALQ-99 - While sustainment and reliability of the 40 year old ALQ-99 systems continues to challenge the DoN (USMC and Navy), we have prioritized NGJ implementation to replace the most stressing frequency coverage first. Navy is developing an interim upgrade solution for the low frequency range transmitter in the Low Band Consolidation (LBC) transmitter set. The LBC is on track to field in the first quarter of Fiscal Year 2020. The LBC does not meet the full requirements of the NGJ Low Band system, however will increase the reliability of the low frequency system.

End of Addendum B

Addendum C
SUMMARY OF CLASS A, B AND C AVIATION-RELATED SAFETY ISSUES

A summary of all Naval Aviation Class A, B and C aviation-related safety issues, including recent mishaps, trends, and analysis from October 2015 through May 24, 2017 follows. The rates presented in the table are based on total mishaps per 100,000 flight hours and include Flight, Flight-Related and Ground mishaps.

Year	Flight Hours	Class A	Class A Rate	Class B	Class B Rate	Class C	Class C Rate
FY16	1,098,519	18	1.64	27	2.46	224	20.39
FY17	689,850	15	2.17	19	2.75	163	23.63

The most recent Fiscal Year 2017 DoN flight Class A mishaps include:

- 26 Apr 2017: (Off the Coast of Guam) MH-60R collided with water on initial takeoff from ship. No injuries.
- 21 Apr 2017: (Philippine Sea) F-18E lost on approach to landing on carrier. Pilot ejected without injury prior to water impact.
- 05 Apr 2017: (Yuma, AZ) CH-53E landed hard and rolled on day training flight. Crew of 5 uninjured.
- 17 Jan 2017: (NAS Meridian, MS) T-45 crashed following a BASH incident on takeoff. Both crewmembers ejected. No fatalities.
- 13 Dec 2016: (Off the Coast of Okinawa, Japan) MV-22B attempted a precautionary emergency landing (PEL) to dry land but crash landed in shallow water. Crew of 5 evacuated with injuries.
- 07 Dec 2016: (Off the Coast of Iwakuni MCAS, Japan) F/A-18C crashed into the water while conducting a night mission. 1 fatality.
- 21 Nov 2016: (Upper Mojave Desert Region) F/A-18F struck a tree while instructor pilot was conducting a currency flight event. Returned to base safely. No injuries.
- 09 Nov 2016: (Off the Coast of San Diego) Two F/A-18As were conducting basic flight maneuvers and had a mid-air collision. 1 aircraft crashed in the water. Pilot ejected successfully. 1 aircraft landed with significant damage
- 27 Oct 2016: (MCAS Beaufort, SC) F/A-35B had an inflight weapons bay fire followed by an uneventful landing. No injuries.
- 25 Oct 2016: (Twentynine Palms, CA) F/A-18C crashed on final approach. Pilot ejected successfully. No injuries.

- 20 Oct 2016: (Yuma, AZ) CH-53E main rotor contacted building causing damage to the aircraft.
- 13 Oct 2016: (Tinker AFB, OK) E-6B #2 engine sustained compressor blade damage due to bird ingestion. Aircraft landed safely. No injuries.

There are three recent FY 2017 DoN Class A aviation ground operations mishaps (AGM):

- 19 January 2017: (NAS Norfolk, VA) Three E-2C aircraft damaged in an engine oil related event. (AGM)
- 18 December 0216: (Kadena Air Force Base, Japan) Tow bar separation resulted in aircraft/tow collision with damage to nose gear and lower fuselage of P-8A. (AGM)
- 16 December 2016: (NAS Whidbey Island, WA) Canopy on EA-18G exploded/jettisoned resulting in severe injuries to two personnel. (AGM)

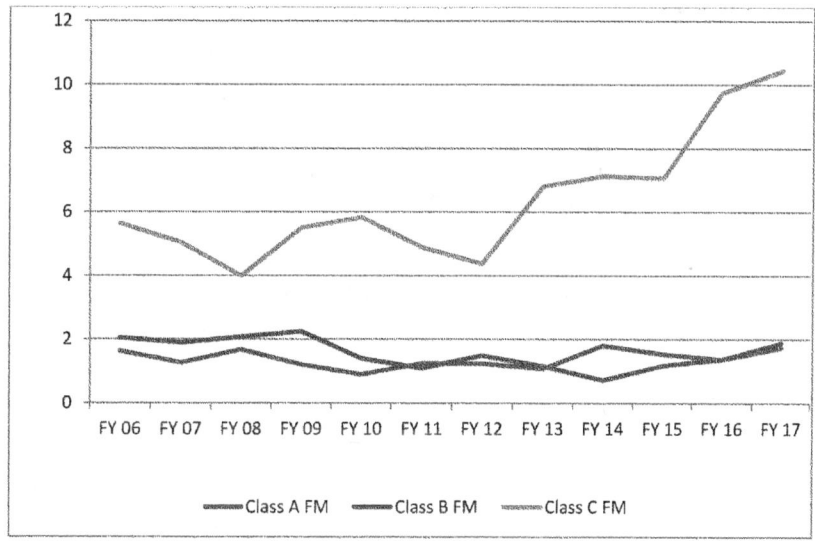

**DoN Historical Mishap Rate Trend per 100K Flight Hours per Mishap Class
(As of 24 May, 2017)**

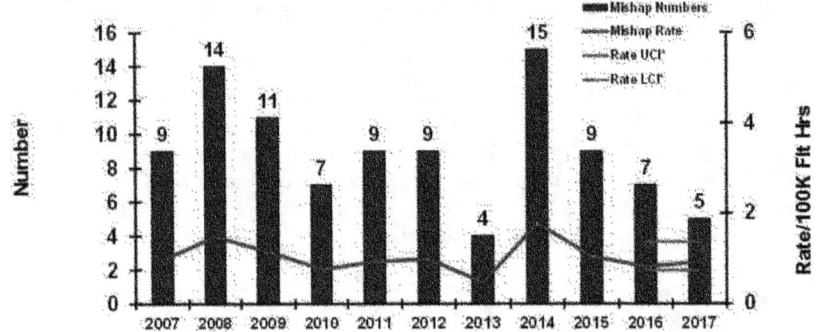

Class A Manned Flight MISHAP Historical Data for U.S. Navy

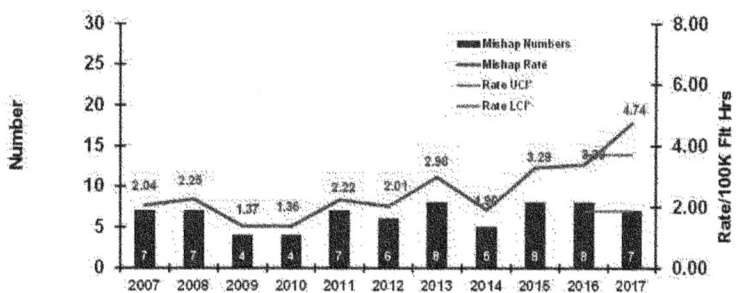

Class A Manned Flight MISHAP Historical Data for U.S. Marine Corps

UCI = Upper Confidence Interval   LCI = Lower Confidence Interval
Rate values above the UCI or below the LCI infer a statistically significant change is probable. This is only an indicator. Significance cannot be determined until end-of-year. Values between the UCI and LCI infer that nothing significant has occurred to increase or decrease mishap rate.

End of Addendum C

Vice Admiral Paul A. Grosklags
Commander
Naval Air Systems Command

Vice Adm. Paul Grosklags is a native of DeKalb, Illinois. He graduated from the U.S. Naval Academy in 1982, is a graduate of the U.S. Naval Test Pilot School Class '99, and holds a Master of Science in Aeronautical Engineering from the Naval Postgraduate School.

After being designated a naval aviator in October 1983, he immediately reported to Training Squadron (VT) 3 at North Whiting Field in Milton, Florida, as a T-34C flight instructor.

Grosklags served operational tours with Helicopter Antisubmarine Squadrons (HS) 34 and 42, where he flew the SH-2F and SH-60B, respectively. Grosklags made multiple deployments with the USS John Hancock (DD 981), USS Donald B. Beary (FF 1085), USS Comte de Grasse (DD 974) and USS Leyte Gulf (CG 55). He later served as both executive and commanding officer of Helicopter Training Squadron (HT) 18.

Grosklags' acquisition tours include engineering test pilot and assignments as MH-60R assistant program manager for systems engineering, H-60 assistant program manager for test and evaluation, MH-60R deputy program manager and ultimately as program manager for Multi-Mission Helicopters (PMA-299), during which time the MH-60R was successfully introduced to the fleet. Grosklags also served as operations officer and subsequently as deputy program executive officer for Air Anti-Submarine Warfare, Assault and Special Mission Programs (PEO(A)).

Grosklags has served flag tours as commander, Fleet Readiness Centers and Naval Air Systems Command (NAVAIR); assistant commander for Logistics and Industrial Operation, NAVAIR, vice commander, PEO(A) and principal military deputy for the Assistant Secretary of the Navy (Research, Development & Acquisition). In October 2015, he assumed responsibilities as Commander, Naval Air Systems Command.

He has more than 5,000 military flight hours in numerous types of rotary and fixed-wing aircraft. Grosklags is a proud but humble co-owner of the Green Bay Packers and works weekends providing free labor on his wife's farm.

Updated: 26 October 2015

Lieutenant General Jon M. Davis
Deputy Commandant for Aviation

Lieutenant General Jon M. Davis assumed his current position as the Deputy Commandant for Aviation, Headquarters Marine Corps in June 2014. Commissioned in May 1980 through the PLC Program, LtGen Davis completed the Basic School in August 1980, and then reported for flight training. Upon receiving his wings in September of 1982, he was selected to fly the AV-8A Harrier.

He reported to VMAT-203 in October 1982, completed Harrier training and reported to VMA-231 in 1983 where he deployed aboard the USS Inchon. In 1985 he transferred to VMAT-203 serving as an instructor pilot. In 1986 he attended the WTI course at MAWTS-1. In 1987 he transferred to VMA-223 serving as the "Bulldogs" WTI and operations officer. From 1988 to 1991 he served as an exchange officer with the Royal Air Force. After training in the United Kingdom, he deployed to Gutersloh, Germany for duty as a GR-5/7 attack pilot with 3(F) squadron. From 1991 to 1994 he served as an instructor at MAWTS-1 in Yuma, AZ. From 1998 to 2000 he commanded VMA-223. During his tour, VMA-223 won the CNO Safety Award and the Sanderson Trophy two years in a row, and exceeded 40,000 hours of mishap free operations. After completing the Executive Helicopter Familiarization Course at HT-18 in Pensacola in 2003, he was assigned to MAWTS-1 where he served as Executive Officer and from 2004 to 2006 as Commanding Officer. From 2006 to 2008 he served as the Deputy Commander Joint Functional Component Command -- Network Warfare at Fort Meade, Maryland. He commanded the 2nd Marine Aircraft Wing from July 2010 to May 2012. From May 2012 to June 2014, he served as the Deputy Commander, United States Cyber Command.

His staff billets include a two year tour as a member of the 31st Commandant's Staff Group, and two years as the Junior Military Assistant to the Deputy Secretary of Defense. In 2003, he served as an Assistant Operations Officer on the 3rd Marine Air Wing staff in Kuwait during Operation Iraqi Freedom. In 2004, he served in Iraq as the Officer in Charge of the 3d Marine Aircraft Red Team. He served as the Deputy Assistant Commandant for Aviation from 2008 to 2010. In the course of his career he has flown over 4,500 mishap free hours in the AV-8, F-5 and FA-18 and as a co-pilot in every type model series tilt-rotor, rotary winged and air refueler aircraft in the USMC inventory.

LtGen Davis graduated with honors from The Basic School and was a Distinguished Graduate of the Marine Corps Command and Staff College. He is a graduate of the Tactical Air Control Party Course, Amphibious Warfare School, Marine Aviation Weapons and Tactics Instructor Course (WTI), The School of Advanced Warfighting (SAW), and Johns Hopkins School of Advanced International Studies (SAIS). He holds a Bachelors of Science from Allegheny College, a Masters of Science from Marine Corps University and a Masters of International Public Policy from Johns Hopkins.

His personal decorations include the National Intelligence Distinguished Service Medal, the Defense Superior Service Medal (two awards), the Legion of Merit (two awards), Meritorious Service Medal (three awards), Navy Commendation (three awards) as well as other campaign and service awards.

Rear Admiral DeWolfe Miller, III
Director, Air Warfare (OPNAV N98)

Rear Adm. DeWolfe Miller hails from York, Pennsylvania, and graduated from the U.S. Naval Academy in 1981. He holds a Master of Science in National Resource Strategy from the National Defense University, is a national security management fellow of the Maxwell School of Citizenship and Public Affairs, Syracuse University and is a graduate of the Navy's Nuclear Power Program.

Miller's command tours include Strike Fighter Squadron (VFA) 34, USS Nashville (LPD 13), USS George H.W. Bush (CVN 77) and as a flag officer, Carrier Strike Group (CSG) 2 providing support to maritime security operations and combat operations for Operations Enduring Freedom and Iraqi Resolve.

Miller's operational tours began after earning his wings of gold in 1983 as a flight instructor with Training Squadron (VT) 19 in Meridian, Mississippi, followed by his first fleet assignment with Attack Squadron (VA) 56, flying the A-7E aboard USS Midway (CV 41) in Yokosuka, Japan. After transitioning to the FA-18 in 1986, subsequent fleet tours included Strike Fighter Squadron (VFA) 25 on USS Constellation (CV 64), department head tour with VFA-131 aboard USS Dwight D. Eisenhower (CVN 69) and executive officer of USS Carl Vinson (CVN 70).

Miller's shore tours include FA-18 test director at Air Test and Evaluation Squadron (VX) 5 in China Lake, California; special aviation programs analyst on the staff of the chief of naval operations (N80); executive officer of Strike Fighter Weapons School Atlantic; deputy director of naval operations at the Combined Air Operations Center during Operation Allied Force; special assistant for Research and Development, Science and Technology and Operational Testing in the Office of Legislative Affairs for the Secretary of Defense; Aircraft Carrier Requirements officer for Commander, Naval Air Forces; and flag officer tours as director, intelligence, surveillance, and reconnaissance capabilities division and assistant deputy chief of naval operations for warfare systems, both in the Office of Chief of Naval Operations.

His personal decorations include the Defense Superior Service Medal, Legion of Merit, Bronze Star, Meritorious Service Medal, Air Medal, Navy and Marine Corps Commendation Medal, Navy and Marine Corps Achievement Medal and various campaign, unit and service awards. He has accumulated more than 4,000 mishap-free flight hours and 877 carrier-arrested landings.

Updated: 17 May 2016

NOT FOR PUBLICATION UNTIL RELEASED BY
HOUSE ARMED SERVICES COMMITTEE
SUBCOMMITTEE ON TACTICAL AIR AND LAND FORCES
UNITED STATES HOUSE OF REPRESENTATIVES

DEPARTMENT OF THE AIR FORCE

PRESENTATION TO THE
HOUSE ARMED SERVICES COMMITTEE
SUBCOMMITTEE ON TACTICAL AIR AND LAND FORCES
UNITED STATES HOUSE OF REPRESENTATIVE

HEARING DATE/TIME: June 7th, 2017, 3:30pm

SUBJECT: Air Force, Force Structure and Modernization Programs

STATEMENT OF:
    Lt. Gen. Arnold W. Bunch, Jr. USAF
    Military Deputy, Office of the Assistant Secretary
    Of the Air Force (Acquisition)

    Lt. Gen. Jerry "JD" Harris Jr., USAF
    Deputy Chief of Staff
    (Strategic, Plans, Programs and Requirements)

NOT FOR PUBLICATION UNTIL RELEASED BY
HOUSE ARMED SERVICES COMMITTEE
SUBCOMMITTEE ON TACTICAL AIR AND LAND FORCES
UNITED STATES HOUSE OF REPRESENTATIVES

INTRODUCTION

Chairman Turner, Ranking Member Tsongas and distinguished members of the Tactical Land Forces Subcommittee, thank you for the opportunity to provide an update on the United States Air Force Modernization programs and Force Structure. For the past 70 years, from the evolution of the jet aircraft to the advent of the ICBM, satellite-guided bombs, and remotely piloted aircraft, the Air Force has been breaking barriers as a member of the finest joint warfighting team on the planet. Today's demand for Air Force capabilities continues to grow as Airmen provide America with unmatched **Global Vigilance**, **Global Reach** and **Global Power**.

In, through and from air, space, and cyber, the fabric of our Air Force weaves multi-domain effects and provides joint warfighters the blanket of protection and ability to power project America's full range of combat capabilities…we're 'Always There'. But, in a world of increasing threats, ever-improving adversaries, and a persistent war against violent extremism, there is a greater disparity than ever before between commitments and the resources necessary to provide unmatched Global Vigilance, Global Reach and Global Power. We are supporting Combatant Commander requirements in response to growing challenges from Russia, China, North Korea and Iran, in addition to the ever present counterterrorism mission in the Middle East and around the world.

While our forces have been heavily engaged in deterring or addressing these operational challenges, our adversaries have taken the opportunity to invest in and advance their own capabilities. To address ever narrowing capability gaps, the Air Force needs your support in the form of, steady and predictable appropriations that fulfill our annual budget requests. Budget levels under the current Budget Control Act restrictions will force the Air Force to continue making unacceptable tradeoffs between force structure, readiness, and modernization. With your support of our FY 2018 budget request, the Air Force can invest in critical capabilities and modernization programs while sustaining capacity and recovering readiness to ensure the joint force can deter, deny and decisively defeat any enemy that threatens the United States or our national interests.

We are committed to providing the most effective bomber, robust tanker, and dominant fighter force to the nation. That is why our top three acquisition priorities in our FY18 Budget Request remain the B-21 Bomber, the KC-46A aerial tanker, and the F-35A Joint Strike Fighter.

ALWAYS THERE Your Air Force relentlessly provides **Global Vigilance, Global Reach**, and **Global Power** for the nation…we're always in demand…and we're always there.

Stitched together, the fabric of our Air Force weaves multi-domain effects and provides U.S. servicemen and women the blanket of protection and the ability to power project America's full range of combat capabilities. Make no mistake, your Air Force is always there.

READINESS IN A CHANGING WORLD Being "always there" comes at a cost to our Airmen, equipment, and infrastructure, and we are now at a decision point. Sustained global commitments and funding constraints have affected capacity and capability for a full-spectrum fight against a near-peer adversary. In 2013, sequestration forced hard decisions that sacrificed the readiness and size of the Total Force in order to ensure our technological superiority against future adversaries. In the FY16 and FY17 budgets, we made the necessary adjustments to balance near-term readiness with future modernization, but due to continuous combat operations, reduced manpower, an aging fleet, and inconsistent funding our readiness has suffered.

In a world of increasing threats, stronger adversaries and a persistent war against violent extremism, there is a greater disparity between commitments and the resources necessary to achieve our national security objectives. Instead of rebuilding readiness for near-peer conflicts, your Air Force is globally engaged in operations against lesser-equipped, but still highly lethal and adaptive enemies. Airmen serve at home and abroad to underpin joint force success but it comes at the expense of full-spectrum readiness.

The first step to regain full-spectrum readiness is to rebuild our Operational Training Infrastructure. This includes not only live, virtual and constructive environments, but also the ranges and space necessary to train against high-end threat systems in a multi-domain environment. Once established, our 4th and 5th generation fighter units need relief from current tasking against low-end adversaries in order to train for emerging threats. We prioritized this initiative by creating a directorate on the Air Staff dedicated solely to this monumental effort. We took the first step, but the complexity of linking all of the systems needed for tomorrow's fight and deconflicting training requires both manpower and finances.

Your Air Force needs permanent relief from the current BCA caps, sufficient funding, flexible execution authority, and manpower to recover full-spectrum readiness. We will continue to do all we can to innovate, transform, and improve how we maximize our resources.

PEOPLE Airmen are our greatest resource and our Air Force needs to increase end strength to meet national security requirements. Manpower shortfalls in key areas remain the number one issue limiting readiness and is our top priority as we rebuild squadrons across the Air Force. At the start of 2016, our end strength stood at 311,000 active duty Airmen, down from more than 500,000 during Desert Storm—a 38 percent decrease. Though we appreciate your support to build the force up to about 321,000 in 2017, we will still be stretched to address national security requirements.

To improve readiness and attain manning levels matching our mission requirements, we worked with the Secretary of Defense to address personnel shortages in the FY 2018 President's

Budget to include an increase in our Active Duty, Guard, and Reserve end strength. Our Total Force model (incorporating our Active Duty, Guard, Reserve, civilians, and our contracted capabilities) not only recognizes the value of an integrated team, but helps guarantee today's and tomorrow's capability. We will develop plans to address shortfalls in a number of key areas, including critical career fields such as aircraft maintenance, pilots, NC3, intelligence, cyber, and battlefield Airmen.

As a Service, we face an aircrew shortage crisis across all disciplines. Your Air Force has the world's finest aircrew who enable an incomparable duality of global mobility and combat lethality. In the aircraft maintenance field, we were short approximately 3,400 aircraft maintainers at the close of 2016. Because of this shortage, we cannot generate the sorties needed for our aircrews. As airlines continue hiring at unprecedented rates, they draw away our experienced pilots. Without a healthy pool of pilots, we risk the ability to provide airpower to the nation.

Pilots are strategic national assets and the pilot crisis extends beyond the Air Force and military. It is a national problem which requires senior-level attention in Congress, the Commercial Industry, and the DoD. To address this national challenge, since 2014 the 'Air Force -Airline Collaboration', formally known as the National Pilot Sourcing Forum has increased efforts to effectively utilize and train an adequate number of pilots to meet our nation's pilot demand signal.

However, pilot retention has declined for five straight years. Today the Air Force has a rated manpower shortfall of approximately 1,550 pilots across the Total Force. This shortfall is most pronounced in our regular Air Force fighter community which is short more than 950 pilots. We are grateful for your support to increase the pilot bonus, and we will continue to ensure our retention programs are appropriately sized and utilized. Your Air Force will utilize the new FY17 NDAA Aviation Bonus authority ($35K per year maximum) and implement a tiered-model using a directed business case model to identify areas of greatest need.

Retaining our pilot force goes beyond financial incentives...it is about culture. Your Air Force is implementing many non-monetary efforts to reinvent the culture and improve the quality of life and quality of service for our Airmen. We have reduced additional duties and superfluous training courses, as well as acquired contracted support in fighter squadrons to perform burdensome administrative tasks, enabling our pilots to focus on their primary duty: flying. We have also increased the transparency of the assignment process and increased flexibility to promote family stability. Your Air Force is exploring opportunities to reduce deployment burdens by enabling more Air Reserve Component volunteers for 179/365-day deployments. We must show our Airmen that we are creating a culture that reminds them they serve in something bigger than themselves...defending America.

In addition to retaining our talented personnel, the Air Force must also increase pilot production and absorption while reducing requirements. The increased end-strength provided in the FY17 NDAA will allow us to maximize the training pipeline and fill out under-manned units, which are vital to our recovery. Our fighter pilot production targets have increased 15% (to 335 Total Force pilots) per year while we surge the number of new aircraft maintainers by more than 1,500 per year to better man flying squadrons and reestablish sortie generation rates with a completion target of 3-5 years. However, other options beyond manpower increases exist to season our young pilots while accelerating readiness recovery.

The Air Force is also investigating a new light attack aircraft (OA-X) that may provide opportunities to create a "high/low" mix for combatting low-end threats in more permissive environments. We have invited industry to participate in a demonstration this summer to determine if a business case exists to add a light attack aircraft to our arsenal. A commercial off-the-shelf OA-X could be used to complement, not replace, our current aircraft inventory. This approach could provide more cockpits to absorb and season a greater quantity of fighter pilots and provide 4th and 5th generation aircraft the required training time to prepare for high-end threats and the operational tempo relief to extend their service life.

SAFETY ISSUES Over the past year, the Air Force is experiencing Class A, B and C mishaps at rates which are lower when compared to the previous 10 year average. This lower than average trend has been sustained over the last two years. Of note, in the past year, the Air Force has realized a significant decrease in Class A mishaps involving Remotely Piloted Aircraft (RPA); primarily due to a large increase in MQ-9 flight hours and a decrease in MQ-1 flying hours. Class A safety issues remain: material failure and aircrew error. Over the past year, when Class A mishaps have resulted in the total loss of the aircraft, about 40% involve some type of material failure. Safety issues in Class B mishaps are related to engine failures and wildlife strikes to aircraft. In the past 12 months, the Air Force has experienced over 50 mishaps involving wildlife strikes where the damage to the aircraft exceeded $500,000, the class B threshold.

Regarding physiological incidents, the preponderance of these incidents result in no damage to the aircraft as the aircrew recognize and properly respond to the incident and safely recover the aircraft. Unfortunately, these incidents are not isolated to one aircraft type or to one oxygen delivery system and as a result, there is most likely not one solution. Therefore the Air

Force continues to pursue technology to measure and report oxygen delivery to the pilot, possible containments in the oxygen system, and overall aircrew physiological state.

The goal in the Air Force is to preserve our combat readiness by eliminating mishaps that result in the loss of aircraft or worse, an Airman. To achieve this goal we continue to pursue technological and material solutions, such as the Auto Ground Collision Avoidance Systems, to enhance pilot performance and prevent mishaps. Finally, a large part of our safety program is done through proactive safety programs which identify and address hazards before they ever result in damaging mishaps or injuring. Mitigating hazards before they injure our Airmen or damage and degrade our combat capabilities is fundamental to the Air Force's pro-active mishap prevention program.

FORCE STRUCTURE AND MODERNIZATION Five years ago during a period of severe fiscal constraints, the Air Force rebalanced our fighter force structure using analysis which showed the Air Force could decrease fighter force structure by approximately 100 aircraft if we were willing to accept higher risk. This resulted in the current fighter inventory of approximately 1,000 primary mission aircraft and slightly more than 1,950 total aircraft. The current inventory complies with FY16 NDAA language on the limitation on retirement of Air Force fighter aircraft; however, with today's sustained operational demand for rotational fighter presence, our current 55 combat-coded fighter squadrons do not allow for enough time at home station to train pilots and maintain aircraft to achieve the full spectrum readiness necessary to meet the requirements set forth in the Defense Planning Guidance.

We need to regrow our current fighter force, both in quantity of fighter squardons and fighter aircraft, across our Active, Guard, and Reserve components. The Fiscal Year 2018 President's Budget begins to address this need by retaining 55 combat squadrons through 2030 and laying the foundation for a fighter force recapitalization. This balance will continue to evolve as we procure more F-35 aircraft and develop Penetrating Counterair (PCA) capability to modernize our fighter force.

However, we are faced with more than just a fleet capacity challenge. Your Air Force's ability to ensure the freedom from attack, freedom to attack, and freedom to maneuver that we provide to the Joint warfighter is being challenged by potential adversaries who are developing and implementing advanced Anti-Access / Area Denial (A2/AD) capabilities. Adversary A2/AD technologies continue to advance at a pace where they will soon out-match our current capabilities, and are being proliferated world-wide as demonstrated by the introduction of advanced Surface-to-Air Missiles in Syria. Modernizing our fleet to address this shrinking gap in capability is one of our top priorities.

Recent fiscal constraints forced your Air Force to make difficult choices in regards to readiness and modernization. With relief from the current BCA caps, we can address both readiness through increased force structure and modernization of the fleet. This relief will allow the Air Force to continue to develop and procure new advanced systems like the F-35A, the B-21, and PCA to address the highly contested threat environment while also modernizing our legacy fleet to ensure these aircraft remain relevant in the contested threat environment.

The Air Force's major modernization focus today is the F-35A, which is the centerpiece of our future fighter precision attack capability. Its primary missions will include Air Interdiction, Offensive and Defensive Counter Air, Close Air Support, Strategic Attack, Suppression of Enemy Air Defenses. The F-35A will also serve as a dual capable aircraft for the U.S. and partner nations. Following the declaration of Initial Operational Capability, Red Flag participation and deployments to Europe, the F-35A has already started proving its mettle. The

Air Force remains on track to field Block 3 capabilities in 2018. This budget request includes $6.3 billion for continued development and procurement of 46 F-35s, but to fill capability and capacity shortfalls, the Air Force needs to increase F-35A procurement to a minimum of 60 aircraft per year as quickly as possible. This must be carefully balanced with the required follow-on modernization effort for the F-35A.

The F-35's follow-on modernization effort centers on the Block 4 upgrade, which is geared toward meeting the estimated threat in the 2025 timeframe and beyond. We cannot emphasize enough how important it is that we fully fund Block 4 to prevent delaying required capabilities for American and Coalition warfighters, including integration of additional weapons and upgrades to the electronic warfare system, data link systems, and radar.

The F-22, currently the only U.S. fighter capable of operating in highly contested environments is also an integral piece of the Air Force's force structure modernization plan. Its stealth, super cruise, integrated avionics and sensors combine to deliver the Raptor's unique capability. We plan to retain the F-22 until the 2060 timeframe, meaning a sustained effort is required to counter advancing threats that specifically target its capabilities. The FY18 budget includes 624.5 million dollars in RDT&E and $398.5 million in procurement towards this goal. New software and hardware in increment 3.2B remains on track to field in FY19and will deliver advanced missile capabilities and improved awareness of ground threats. The FY18 budget also funds the acceleration of the TACLink 16 program, which adds transmit capability for the Raptor—providing situational awareness to all US and coalition fighters through the Link 16 network.

FY18 begins an increase in the Air Force's commitment to fielding a future penetrating counterair capability following the recommendations of the Air Superiority 2030 Enterprise

Capability Collaboration Team. As our adversary capabilities advance, a new PCA capability will play a critical role in targeting and engaging future threats in the most highly contested environments. It will also be instrumental as a node in the larger network, providing data from its sensors to enable complementary weapon systems. This capability will provide the survivability, lethality and persistence to meet emerging worldwide threats across the spectrum of conflict and will be the cornerstone of the Air Force shift from $4^{th}/5^{th}$ generation to a $5^{th}/6^{th}$ generation fleet.

In addition to pursuing new capabilities and modernizing fifth generation fighters, the Air Force also seeks to extend the service life and modernize critical capabilities of key fourth generation aircraft. Doing so will help maintain Service capacity and readiness to meet the needs of today's counterterrorism fight while ramping up the F-35 production line and developing PCA.

The legacy service life extension program (SLEP) will extend the F-16 airframe structural service life from the current 8,000 hours to more than 12,000 hours, adding fifteen to twenty years of service for 300 selected F-16s through an effort budgeted at $350 million dollars. To ensure the F-16's lethality and preminence for homeland defense and current conflicts, we are pursuing an active electronically scanned array (AESA) radar upgrade that offers advanced capabilities and improved reliability and maintainability over current systems. The contract was awarded on 31 May, leading to initial operational capability in the second quarter of 2019 for the Homeland Defense Aerospace Control Alert mission requirement. We are also upgrading the mission computer, display generator, electronic warfare components, and the ALQ-131 self-protection jamming pod, known as the Pod Upgrade Program (PUP) that enables advanced technology jamming techniques.

Along with the F-16, the Air Force expects the F-15E to be an integral part through at least 2040 and we are pursuing a new electronic warfare self-protection suite, the Eagle Passive/Active Warning Survivability System (EPAWSS) for the Strike Eagle fleet. Based on the interim results of a full-scale fatigue test, due to be completed in 2018, no service-life extension programs are currently planned for the F-15E.

We also continue to modernize our F-15C/D fleet with AESA radars, a more capable aircraft mission computer, an infrared frequency targeting sensor and a more robust and powerful data link. To ensure the integrity of the F-15 airframe we are replacing the fuselage longerons starting in FY2018, mitigating risks to F-15 aircrew and ensuring integrity of the aircraft into the next decade. The program is budgeted at $205 million for 235 aircraft. The Air Force anticipates recapitalization of a portion of the F-15C/D fleet in the 2020-2030 timeframe as we balance capability, sustainability and capacity across the fighter force.

This year's budget also provides $17.5 million in investment funding for the A-10 weapon system. Full funding for sustainment and modernization postures your Air Force to keep the full fleet of A-10s relevant until after F-35 Initial Operational Test and Evaluation is complete. It also provides $6 million to begin procurement of hardware under the ADS-B program to meet FAA mandates. Pending IOT&E results, the Air Force is committed to maintaining a minimum of six A-10 combat squadrons flying and contributing to the fight through 2030. Additional A-10 force structure is contingent on future budget levels and force structure requirements.

The Air Force will not be able to rely solely on our current programs and capabilities to ensure readiness to fight the most advanced threats in the future. To that end, we are aggressively pursuing a path toward strategic agility in our capability development. We have reinvigorated development planning (DP) at the enterprise level to build-in agility and formulate truly innovative strategic choices for capability development. Core Development Planning functions include: formulating and evaluating viable future concepts, defining operational trade space, identifying technology shortfalls and Science and Technology needs, and assisting the operations community in refining requirements.

To oversee and direct capability development of the highest priority operational challenges and opportunities, the Air Force established the 3-star Capability Development Council (CDC), chaired by the Vice Chief of Staff of the Air Force, as well as stood up the Strategic Development Planning and Experimentation (SDPE) office to plan, manage, and execute warfighting experimentation campaigns. Experimentation provides the ability to rapidly explore a wide range of innovative materiel and non-materiel solution options. To further these efforts, the Air Force programmed resources starting in Fiscal Year 2017 to conduct concept-driven experimentation campaigns, including funds for prototyping, live and virtual simulations; developing a cadre of expertise, along with the tools to conduct experimentation campaigns.

The Light Attack Experimentation Campaign informs planning and strategic choices in this critical area. The Air Force is experimenting with potential off-the-shelf aircraft as part of a broader assessment into industry's capability, capacity, and interest to provide cost-effective innovative solutions with low procurement, operating and sustainment costs. Since the deployment demand is not expected to decrease, the Air Force must meet capability demands in permissive environments while building and maintaining readiness to meet emerging threats in more contested environments. Aligning capability, capacity and cost with wartime demands is key to meeting Air Force commitments to combatant commanders and effectively using taxpayer resources. Assessing the viability of low operating cost, light attack platforms has the potential to reduce operating costs while still meeting combatant commander needs.

After completing an evaluation of all respondents under the competitive process outlined in the invitation, the Air Force notified invited companies of their selection to participate in the live-fly experiment this summer. We are currently in the process of reaching agreement on Other Transaction Agreements with these companies to outline the details of their participation. This live-fly experiment will assess the capabilities of these off-the-shelf light attack aircraft, which will be flown by Air Force personnel in scenarios designed to highlight aspects of various combat missions, such as close air support, armed reconnaissance, combat search & rescue and strike control and reconnaissance. The experiment will also include the employment of weapons commonly used by other fighter/attack aircraft to demonstrate the capabilities of light attack aircraft for traditional counter-land missions. Results from this experimentation campaign will be used to inform future capability development and investment decisions.

MUNITIONS There is an ever growing demand for the effects airpower brings to the joint force. Within our fiscal boundaries, we have sought to balance the requirement for current munitions with the need to advance capabilities in the same manner we have with our aircraft force structure. However, sustained combat operations, BCA limitations, and support for our coalition partners have negatively impacted these efforts. Absent sustained and increased funding, munition stockpiles will continue to decrease as well as negatively impact readiness and our ability to meet national security objectives in the future.

Historically, munitions funding has been reduced to pay other critical service bills. To resolve this issue, we need increased and sustained funding at our FY 2018 requested levels to send a more consistent demand signal to our industrial base. With the dispensation provided by the Congressional Defense Committees, we were able to utilize the Overseas Contingency Operations funding to replenish munitions with high combat expenditures.

We are currently using legacy munitions on our 5th generation fleet which negates the full advantage these platforms can provide. Investments into programs such as the Small Advanced Capabilities Missile (SACM) and the Stand in Attack Weapon (SiAW) are crucial to realizing the full potential of our next generation of aircraft. The SACM is a smaller, affordable air to air weapon that is required to increase magazine depth and maximize utility of a PCA capability. SiAW is an air-to-surface weapon designed to hold at risk the surface elements that make up the A2AD environment and will be integrated on F-35 and other future platforms like PCA. With your continued help the USAF must continue to invest in and develop advanced munition capabilities such as these to ensure future air superiority for the Joint Force.

INTELLIGENCE, SURVEILLANCE & RECONNAISSANCE (ISR) The RQ-4 Global Hawk provides a continuous, high altitude long endurance all weather, day/night, wide area reconnaissance and surveillance unmanned aircraft system. The Office of Secretary of Defense approved the RQ-4 modernization approach in September 2015 to include the MS-177 sensor integration, a Ground Segment Modification Program and a Communication System Modification Program. The MS-177 development and integration work began in November 2015 and the sensor is on track for Initial Operating Capability (IOC) in First Quarter FY18. The FY18 PB request is for $383.2 million in investment dollars for this program.

The Ground Segment Engineering & Manufacturing Development (EMD) contract was awarded in July 2016. Installation of cockpits at Grand Forks AFB and Beale AFB will begin in First Quarter FY18. The Communication System Modification Program (CSMP) effort is in the Requirements Definition/Market Research phase. The program is finalizing requirements for modernization of Ground and Air Vehicle communications equipment, which will both improve communications capability and alleviate Diminishing Manufacturing Sources (DMS) issues with the equipment. We expect to field the CSMP in the 2022-2025 timeframe.

The funding request for the MQ-9 program in FY18 is $1.1 billion. This program continues to modernize it's fleet and capabilities it provides to Combatant Commanders. It accomplishes this by sustaining the MQ-9 program of record and incorporating planned modernization efforts, while a separate program of record develops and tests those modernizations making them ready for the program at large. This process keeps the MQ-9s current and able to meet Combatant Commanders demands, while keeping an eye on the future for emerging requirements. Such efforts include the new Ground Control Station – Block 50 that is actively being developed, the new DAS-4 sensor package that will fly on the MQ-9 platform and the Extended Range enhancement to the MQ-9 Block 5 aircraft. Additionally, the MQ-9 program is actively engaged in a study to determine the actual service life of the MQ-9 platform. The first phase of that study will be completed in summer FY17, with phase two being completed in FY20. The results of this study will better inform the Air Force's decision on long-term sustainment of the MQ-9 program.

Gorgon Stare has been delivering Wide Area Motion Imagery (WAMI) in support of Operation Freedom Sentinel and Operation Inherent Resolve areas of responsibility since 2012. The Air Force has no plans to fund additional capability at this time but will sustain this MQ-9 podded WAMI capability in its current state. The FY18 request to for $85.6 million in Operation and Maintenance funding for this sustainment effort. The Air Force is migrating its primary ISR Processing, Exploitation and Dissemination (PED) weapon system, the Distributed Common Ground System (DCGS), to an Open Architecture. To support this effort $193.8 million has been requested in the FY18 PB. The previous architecture required 5-7 years of development, test, and fielding per major release. Open Archecture will support software releases in weeks and months. This accelerated development and fielding timeline will enhance our ability to get inside the adversaries decision cycle, enable our ISR analysts to leverage cutting-edge analytic tools, and allow increased access to more intelligence sources and Intelligence Community (IC) capabilities.

MULTI-DOMAIN COMMAND AND CONTROL (MDC2) An MDC2 capability generates effects that present the adversary with multiple dilemmas at an operational tempo that cannot be matched. Your Air Force is focused on creating feasible investment options throughout its BMC2 portfolio that drive towards the attainment of an advanced MDC2 capability for the joint force. At the tactical edge, the AWACS weapon system integrates multi-domain inputs to provide air, land, and sea Battle Management and Command and Control (BMC2). The FY18 PB includes a request for $506.2 million for the AWACS program. To ensure the United States maintains mulit-domain dominance, multiple AWACS modernization activities are underway with the most notable being the upgrade to the Block 40/45 mission system which is the foundation for all future AWACS capability improvements. Additionally, the Air Force is in the midst of accomplishing activities for a follow-on battle management command and control capability, the Advanced Battle Management and Surveillance (ABMS), which is currently provided by the E-3/AWACS fleet. The ABMS system is envisioned to be an evolutionary leap in capability intended to achieve IOC prior to the end of AWACS projected service life in 2035.

The E-8C Joint Surveillance Target Attack Radar System (JSTARS) executes Battle Management and Surveillance of air-to-ground operations, an integral piece to today's fight. $417.2 million has been requested in FY18 for the JSTARS Recapitalization program. Our

JSTARS recapitalization strategy integrates mature sensor, communications and battle management technologies on a business class aircraft; the results should reduce life cycle cost while increasing operational availability and mission system capability. As a service we seek to balance mission capability, risk and cost, and will look for opportunities to accelerate the recapitalization as the program progresses. As the Air Force transitions to JSTARS Recapitilization, we remain sensitive to the critical role JSTARS fills for Combatant Commanders and recognize the demand for this capability will likely not decline. As a result, the Air Force remains committed to delivering JSTARS Recapitalization as soon as possible to avoid a potential capacity gap. The program is currently in source selection, upon contract award, the Air Force will further assess any potential capacity gap.

While the Air Operations Center (AOC) Weapon System (WS) Increment 10.2 is currently in a strategic pause, interoperability with the MDC2 vision is essential to the AOC way ahead. The fielded AOC WS 10.1 legacy system will not be able to support the vision for MDC2 without significant improvement/modernization and the Air Force is committed to fielding a modern architecture for the AOC that enables MDC2. During this strategic pause, the program office is partnering with the Defense Innovation Unit Experimental (DIUx) and the Defense Digital Services (DDS) to explore a pathfinder effort to establish an Agile DevOps pipeline to rapidly deliver capability to a single AOC. This pathfinder will help inform the way forward for modernizing the AOC and providing a system capable of being the foundation of MDC2 operations. The AF has requested $119.7 million in the FY18 PB for the AOC program.

ROTORCRAFT The FY18 PB continues investment in your Air Force's critical rotorcraft modernization programs. The FY18 PB requests $88.21 million for the CV-22 fleet to assist in execution of the National Military Strategy by providing transformational mission capability to special operations forces warfighters. The Air Force continues to make improvements to the CV-22 with modifications designed to improve reliability, survivability and capability. Future efforts will make the CV-22 more cost-effective, while ensuring the viability of its unique long-range payload capacity coupled with vertical take-off and landing capability.

The Air Force is the only Service with a dedicated force organized, trained, and equipped to execute theater-wide Personnel Recovery. The newly designated combat rescue helicopter (CRH) will be specifically equipped to conduct Combat Search and Rescue across the entire spectrum of military operations. Due to the advancing age and current attrition rates of the HH-60G, the Air Force must continue to modify existing HH-60G helicopters while utilizing the Operational Loss Replacement program to meet Combatant Command requirements until we can fully recapitalize with the CRH program. In addition to 112 air vehicles, the CRH program will provides for training devices, support equipment and the necessary post production support to successfully field a replacement for the HH-60G. The AF has fully funded CRH research and development across the FYDP to meet National Military Strategy objectives through Personnel Recovery missions. The FY18 PB requests $76 million and $354.5 million for the HH-60G and CRH programs.

Furthermore, the current UH-1N fleet supports a wide range of missions for 5 major commands. It does not however meet speed, range, payload, or survivability requirements. The risk created by these capability gaps makes replacing the UH-1N a critical priority and a vital element of our nuclear enterprise reform initiative. The FY18 President's Budget requests $108.6 million for the UH-1N Replacement Program across and reflects a full and open competitive procurement, which will integrate non-developmental items into off-the-shelf production helicopters to replace the entire UH-1N fleet.

TRAINERS The FY18 PB continues investment efforts for Air Force trainer platforms, including modernization programs for the T-1, T-6, and T-38 fleets. The T-1A Avionics Modernization Program will modernize the T-1A fleet and address known obsolescence and diminishing manufacturing capability issues. The AF is working to install ADS-B Out across the entire T-6 fleet, modernize the Aircrew Training Device, modify the Canopy Fracture Initiation System, and support engineering change proposals and logistics support. Modifications are also required to sustain and upgrade the T-38C fleet, including Pacer Classic III and avionics upgrade programs, until T-X is delivered. The FY18 PB requests $21.5 million, $38.7 million, and $53.6 million for the T-1, T-6, and T-38 fleets, respectively.

The PB also requests $106 million for the Advanced Pilot Trainer (T-X) program, which will provide student pilots in the Specialized Undergraduate Pilot Training advanced phase and Introduction to Fighter Fundamentals with the skills and competencies required to transition into 4th- and 5th-generation fighter aircraft. This new training capability will enable pilots to receive realistic training in a system similar to fielded fighters. It will replace the existing fleet of 430 T-38C aircraft with 350 aircraft and associated Ground Based Training Systems, ground equipment, spares, and support equipment. The T-X program is currently in source selection and plans to award a contract 1QFY18 to ensure we meet a 2024 Initial Operational Capability and 2034 Full Operational Capability.

SUMMARY

The demand for air, space, and cyber power is growing and our Chief is committed to ensuring that America's Airmen are resourced and trained to fight alongside the Army, Navy, Marines and Coast Guard to meet national security obligations. The Air Force seeks to balance risk across capacity, capability, and readiness to maintain an advantage, however persistently

unstable budgets and fiscal constraints have driven us to postpone several key modernization efforts. These delays created a rapid approaching modernization bow wave that includes programs critical to meet our capacity and capability requirements across all mission areas.
The delays have also opened an opportunity to our competitors to close gaps and negate our traditional advantages.

The result of these changes by the world is a marked decrease in our technological advantage. The Air Force once had a decided advantage across all fronts. Today, the Air Force has some advantage in some technological areas however potential adversaries are nipping at our heals or shoulder to shoulder with us in others. To address the shrinking technology gap, we must modernize and continue to invest in S&T so we can ensure we grow back the technology gap so our most valued treasure – America's sons and daughters – we send into harm's way have a decisive advantage....we do not want a fair fight.

Although we are grateful for the recent fiscal relief, we still face uncertainty. Sustainable funding across multiple fiscal year defense plans is critical to ensure we can meet today's demand for capability and capacity without sacrificing modernization for tomorrow's high-end fight against a full array of potential adversaries.

As critical members of the joint team, the USAF operates in a vast array of domains and prevails in every level of conflict. However, we must remain focused on delivering **Global Vigilance**, **Global Reach** and **Global Power**, through our core missions of Air Superiority, Space Superiority, Global Strike, Rapid Global Mobility, ISR, and C2 to continue to provide our nation with security it enjoys. We look forward to working closely with the committee to ensure the ability to deliver combat air power for America when and where we are needed.

Lieutenant General Arnold W. Bunch, Jr.

Lt. Gen. Arnold W. Bunch, Jr., is the Military Deputy, Office of the Assistant Secretary of the Air Force for Acquisition, the Pentagon, Washington, D.C. He is responsible for research and development, test, production, and modernization of Air Force programs worth more than $32 billion annually.

General Bunch was commissioned in 1984 as a graduate of the U.S. Air Force Academy. He completed undergraduate pilot training in 1985. He completed operational assignments as an instructor, evaluator and aircraft commander for B-52 Stratofortresses. Following graduation from the Air Force Test Pilot School, General Bunch conducted developmental testing in the B-2 Spirit and B-52 and served as an instructor in each. Additionally, he has commanded at the squadron, group and wing levels. Prior to his current assignment, he was the Commander of the Air Force Test Center, headquartered at Edwards Air Force Base, California.

EDUCATION
1984 Bachelor of Science degree in civil engineering, U.S. Air Force Academy, Colorado Springs, Colo.
1991 Squadron Officer School, Maxwell AFB, Ala.
1994 Master of Science degree in mechanical engineering, California State University Fresno
1996 Army Command and General Staff College, Fort Leavenworth, Kan.
2000 Master of Science degree in national security strategy, National War College, Fort Lesley J. McNair, Washington, D.C.

ASSIGNMENTS
1. July 1984 - July 1985, Student, undergraduate pilot training, Columbus Air Force Base, Miss.
2. August 1985 - December 1985, Student, B-52 Combat Crew Training School, Castle AFB, Calif.
3. January 1986 - June 1990, Standardization and Evaluation Instructor Aircraft Commander, 325th Bomb Squadron, Fairchild AFB, Wash.
4. July 1990 - June 1991, Student, USAF Test Pilot School, Edwards AFB, Calif.
5. July 1991 - June 1992, Test Pilot, 6512th Test Squadron, Edwards AFB, Calif.
6. July 1992 - June 1995, Test Pilot, 420th Test Squadron, Edwards AFB, Calif.
7. June 1995 - June 1996, Student, Army Command and General Staff College, Fort Leavenworth, Kan.
8. July 1996 - July 1999, Chief, B-1 Test and Evaluation, B-1 System Program Office, Wright-Patterson AFB, Ohio
9. August 1999 - June 2000, Student, National War College, Fort Lesley J. McNair, Washington, D.C.
10. June 2000 - July 2002, Commander, 419th Flight Test Squadron, Edwards AFB, Calif.
11. August 2002 - April 2003, Chief, Senior Officer Management, Air Force Materiel Command, Wright-Patterson AFB, Ohio
12. April 2003 - June 2004, Deputy Chief, Combat Forces Division, the Pentagon, Washington, D.C.
13. June 2004 - January 2006, Director, Munitions Directorate, Air Force Research Laboratory, Eglin AFB, Fla.
14. January 2006 - May 2008, Commander, 412th Test Wing, Edwards AFB, Calif.
15. June 2008 - March 2010, Vice Commander, Air Armament Center, Eglin AFB, Fla.
16. March 2010 - June 2011, Director and Program Executive Officer for the Fighters and Bombers Directorate, Aeronautical Systems Center, Wright-Patterson AFB, Ohio
17. June 2011 - June 2012, Commander, Air Force Security Assistance Center, AFMC, Wright-Patterson AFB, Ohio
18. June 2012 - June 2015, Commander, Air Force Test Center, Edwards AFB, Calif.
19. June 2015 - present, Military Deputy, Office of the Assistant Secretary of the Air Force (Acquisition)

FLIGHT INFORMATION
Rating: command pilot
Flight hours: more than 2,500 hours
Aircraft flown: B-52, B-2, KC-135, F-16, T-38 and others

MAJOR AWARDS AND DECORATIONS
Legion of Merit with two oak leaf clusters
Meritorious Service Medal with five oak leaf clusters
Aerial Achievement Medal with oak leaf cluster
Air Force Commendation Medal
Air Force Achievement Medal
Combat Readiness Medal
National Defense Service Medal with oak leaf cluster
Global War on Terrorism Service Medal

EFFECTIVE DATES OF PROMOTION
Second Lieutenant May 30, 1984
First Lieutenant May 30, 1986
Captain May 30, 1988
Major Dec. 1, 1995
Lieutenant Colonel Sept. 1, 1998
Colonel June 1, 2004
Brigadier General May 7, 2010
Major General Aug. 23, 2013
Lieutenant General June 24, 2015

(Current as of June 2015)

Lieutenant General Jerry D. Harris, Jr.

Lt. Gen. Jerry Harris is Deputy Chief of Staff for Strategic Plans and Requirements, Headquarters U.S. Air Force, Washington, D.C. In support of the Chief of Staff and Secretary of the Air Force, General Harris leads the development and integration of the Air Force strategy, long-range plans and operational capabilities-based requirements. He directs and coordinates activities ensuring the Air Force builds and employs effective air, space and cyber forces to achieve national defense objectives.

General Harris entered the Air Force in 1985 as a graduate of the ROTC program at Washington State University. He has served as a flight commander, operations officer, weapons officer and inspector general. The general served on the staffs of two numbered Air Forces and one major command, all in operations. He has also served as the Combined Air and Space Operations Center Battle Director for operations Iraqi Freedom and Enduring Freedom. General Harris has commanded at squadron, group and wing levels. Prior to his current assignment, General Harris was the Vice Commander, Air Combat Command, Langley Air Force Base, Virginia, responsible for organizing, training, equipping and maintaining combat-ready forces for rapid deployment and employment while ensuring strategic air defense forces are ready to meet the challenges of peace time air sovereignty and wartime defense.

General Harris is a command pilot with more than 3,100 flying hours in the F-16.

EDUCATION
1985 Bachelor of Science in Mechanical Engineering, Washington State University
1992 Squadron Officer School, Maxwell AFB, Ala
1997 Air Command and Staff College, Maxwell AFB, Ala.
1997 Master of Science in Aeronautical Science Technology, Embry-Riddle Aeronautical University, Daytona Beach, Fla.
1998 School of Advanced Airpower Studies, Maxwell AFB, Ala.
1998 Master of Science in Airpower Art and Science, School of Advanced Airpower Studies, Maxwell AFB, Ala.
1998 Armed Forces Staff College, Norfolk, Va.
2001 Air War College, by correspondence
2006 National Defense College, New Delhi, India
2011 Capstone General and Flag Officer Course, National Defense University, Washington, D.C.

ASSIGNMENTS
1. February 1986 - January 1987, Student, undergraduate pilot training, Williams AFB, Ariz.
2. January 1987 - April 1987, Student, AT-38B lead-in fighter training, Holloman AFB, N.M.
3. April 1987 - December 1987, Student, F-16 B-Course, MacDill AFB, Fla.
4. December 1987 - July 1989, Chief, Current Operations Division; Squadron Assistant Programmer; Training Officer; and Mobility Officer, Nellis AFB, Nev.
5. August 1989 - January 1992, Chief of Weapons and Tactics and Air-To-Surface Weapons Officer, Moody AFB, Ga.
6. January 1992 - February 1992, Student, Squadron Officer School, Maxwell AFB, Ala.
7. February 1992 - March 1994, Chief of Mid-range Programming and Student, Fighter Weapons School, Luke AFB, Ariz.
8. March 1994 - June 1996, Weapons and Tactics Flight Commander; Chief of Wing Weapons; and Chief of Squadron Weapons, Eielson AFB, Alaska
9. July 1996 - July 1998, Student, School of Advanced Airpower Studies and Air Command and Staff College, Maxwell AFB, Ala.
10. July 1998 - September 1998, Student, Armed Forces Staff College, Norfolk, Va.
11. September 1998 - March 1999, NATO Joint Staff Officer, Long-range Plans, Plans and Policy,

Headquarters, Southern Region Air Command, Naples, Italy
12. March 1999 - August 2000, Chief of Strategy, Crisis Action Group, Headquarters Southern Region Air Command, Naples, Italy
13. September 2000 - January 2001, Student, F-16 requalification, Luke AFB, Ariz.
14. January 2001 - February 2003, Operations Officer and Chief of Standardization and Evaluation, 20th Operations Group; and assistant Director of Operations, 79th Fighter Squadron, Shaw AFB, S.C.
15. March 2003 - February 2005, Commander, 79th Fighter Squadron, Shaw AFB S.C.
16. March 2005 - July 2005, Staff Director and Inspector General, 20th Fighter Wing, Shaw AFB S.C.
17. July 2005 - December 2005, Deputy Commander, 20th Operations Group, Shaw AFB S.C.
18. January 2006 - January 2007, Student, National Defense College, New Delhi, India
19. January 2007 - July 2008, Commander, 505th Training Group, Hurlburt Field, Fla.
20. July 2008 - November 2008, Director of Air, Space and Information Operations, 13th Air Force, Hickam AFB, Hawaii
21. November 2008 - September 2009, Commander, 8th Fighter Wing, Kunsan Air Base, South Korea
22. September 2009 - September 2010, Assistant Director of Operations, Plans, Requirements and Programs, Headquarters Pacific Air Forces, Hickam AFB, Hawaii
23. September 2010 - September 2012, Commander, 56th Fighter Wing, Luke AFB, Ariz.
24. September 2012 - March 2014, Vice Commander, 5th Air Force, Yokota Air Base, Japan
25. March 2014 - April 2015, Director of Programs, Office of the Deputy Chief of Staff for Strategic Plans and Programs, Headquarters U.S. Air Force, Washington, D.C.
26. April 2015 - February 2017, Vice Commander, Air Combat Command, Joint Base Langley-Eustis, Va.
27. February 2017 - Present, Deputy Chief of Staff for Strategic Plans, Programs, and Requirements, Headquarters U.S. Air Force, Washington, D.C.

SUMMARY OF JOINT ASSIGNMENTS
September 1998 - August 2000, NATO Joint Staff Officer, Long-range Plans, Plans and Policy; and Chief of Strategy, Crisis Action Group, Headquarters Southern Region Air Command, Naples Italy, as a major

FLIGHT INFORMATION
Rating: command pilot
Flight hours: more than 3,300
Aircraft flown: F-16, T-37, T-38, Mig-29 and Mig-21

AWARDS AND DECORATIONS
Distinguished Service Medal
Legion of Merit with two oak leaf clusters
Defense Meritorious Service Medal
Meritorious Service Medal with two oak leaf clusters
Air Medal with three oak leaf clusters
Aerial Achievement Medal
Air Force Commendation Medal with two oak leaf clusters
Joint Service Achievement Medal
National Defense Service Medal with bronze star
Southwest Asia Service Medal with three bronze stars
Kuwait Liberation Medal (Kingdom of Saudi Arabia)
Kuwait Liberation Medal (government of Kuwait)

EFFECTIVE DATES OF PROMOTION
Second Lieutenant May 11, 1985
First Lieutenant Sept. 1, 1987
Captain Sept. 1, 1989

Major Sept 1, 1995
Lieutenant Colonel April 1, 2000
Colonel Jan. 1, 2006
Brigadier General Nov. 3, 2010
Major General June 27, 2014
Lieutenant General Feb. 22, 2017

(Current as of February 2017)

WITNESS RESPONSES TO QUESTIONS ASKED DURING THE HEARING

June 7, 2017

RESPONSES TO QUESTIONS SUBMITTED BY MR. LANGEVIN

Admiral GROSKLAGS and Admiral MILLER. To date, no Unmanned Aircraft Systems (UAS) developed by the Department of the Navy have been used in support of Humanitarian Assistance and Disaster Relief (HA/DR) missions. However, the Navy and Marine Corps have led, and participated in, Joint Task Forces supported by UAS owned and operated by other organizations. The importance of capabilities provided by unmanned systems when planning a coordinated response is well recognized by military leadership, Government Organizations, Non-Government Organizations (NGO), Private Volunteer Organizations (PVO), and other international responders. The earliest and best known instance of military unmanned aircraft systems (UAS) used in a HA/DR role was the use of a RQ–1 Predator in support of Operation Unified Response after the 2010 earthquake in Haiti. After Hurricane Matthew in 2016, Guardian UA, owned and operated by the Customs and Border Patrol, provided surveillance and reconnaissance support to the Joint Task Force Commander supervising hurricane relief efforts in Haiti. The Department of the Navy, in partnership with commercial companies, is exploring development of unmanned cargo delivery systems with varying load capacities to support our warfighters. While the primary purpose is supporting warfighters in combat, there is potential use across the broader military mission spectrum. As the Navy continues to mature our operational concepts for UAS systems, we will utilize UAS across our entire mission set, including HA/DR.   [See page 19.]

General DAVIS. [The information referred to was not available at the time of printing.]   [See page 19.]

General BUNCH and General HARRIS. The Air Force provided MQ–1 Predator and RQ–4 Global Hawk to Combatant Commanders globally to support of multiple humanitarian efforts. In January 2010, the Southern Command deployed the MQ–1 aircraft to Puerto Rico in support of Haiti relief efforts following the January 12, 2010 earthquake. The MQ–1 provided Full-Motion Video (FMV) to identify heavily impacted areas and relay timely information. In March 2011, Pacific Command tasked RQ–4 aircraft to fly over Japan and provide overhead imagery that directly impacted relief efforts. Also, in October 2016 after Hurricane Matthew struck Haiti, Southern Command employed an RQ–4 imaging facilities and infrastructure to guide U.S. efforts providing aid.

Overall, the DOD uses Remotely Piloted Aircraft (RPA) in support of humanitarian missions abroad in the same manner as it uses manned aircraft. These systems, specifically the MQ–9 Reaper and the RQ–4, are both on a modernization timeline that will grow and sustain the sensor capabilities, aircraft loiter times, and utilization across multiple mission areas to include humanitarian support. At this timethe Air Forces sees that it is doing everything appropriate to ensure RPAs are available to meet a wide range of Combatant Commander needs well into the future. [See page 19.]

General HARRIS. Two Air Force lines of effort address cross-platform communications in an Anti-Access/Area Denial environment. In the near term, driven by the Air Superiority 2030 Flight Plan, the Air Force is conducting small scale experimentation campaigns to reduce risk and to expedite fielding of Advanced Tactical Datalinks and enhancements to existing tactical datalinks. These experiments are demonstrating correlation/fusion of data from multiple sources, including intelligence sources and 5th generation fighters. Also, the Agile Communication Capabilities Based Assessment (CBA) is defining communication gaps that the Air Force must mitigate in A2/AD environments in the 2030+ timeframe. The outcome of each of these efforts will inform the path forward for communications capabilities that enable interoperability across the A2/AD environment. Currently the F–22 has a funded program. TACLink 16, that will add Link 16 transmit capability. This will allow the two aircraft to communicate during operations. This program will begin fielding in FY 2021.   [See page 19.]

RESPONSES TO QUESTIONS SUBMITTED BY MS. McSALLY

Admiral GROSKLAGS and Admiral MILLER. The Department of the Navy does not operate the A-10 aircraft. I recommend this request be forwarded to the U.S. Air Force team for official response. [See page 19.]

General DAVIS. [The information referred to was not available at the time of printing.] [See page 19.]

General BUNCH and General HARRIS. The Air Force has a requirement to retain the A-10 for the foreseeable future and is planning to retain the entire 283 aircraft A-10 fleet at a minimum until 2021. Future updates to the DPG as well as completion of the final F-35 Initial Operational Test and Evaluation (IOT&E) Block 3F report may inform a decision to retain fewer than 283 A-10s beyond 2021.

However, beyond 2021, the Air Force can only afford to retain 173 of the 283 A-10s at current budget levels. An additional Total Obligation Authority (TOA) is needed in 2018 and beyond to keep additional aircraft. Including new A-10 wings in the FY18 UPL shows the Air Force requires more than 173 of our A-10s through 2030, but that we cannot fund them internally based on priorities and tough planning choices.

283 A-10s provide two training units, 9 combat coded squadrons, and the associated test and weapons school assets needed. 173 A-10s provide one large training unit, 6 combat squadrons, and the associated additional assets.

Future Options for the A-10 Fleet

283 A-10s—Purchase additional wings: 283 A-10s can only be retained beyond 2021 with approximately $1.4B of additional Air Force TOA in the FYDP. The significant cost of retaining the additional 110 A-10s consists of both buying new wings ($954M total, $380M in FYDP), and sustaining the aircraft operations and maintenance ($500M per year).

Initiating this acquisition program in FY18 would establish an ability to buy additional wings in the future with corresponding Air Force TOA increases. With funding in place, it would take approximately one year to get on contract, with delivery of the first wing as early as FY22. By FY22 approximately 51 A-10s will be grounded due to their wings, depending on depot capacity to overhaul existing A-10 wings. Total cost for 110 new wings is approximately $954 million, with last delivery approximately 11 years after initial award (2030 completion with funding initiated in FY18).

The FY18 Unfunded Priority List (UPL) includes $103M that, in addition to the $20M Congressional add from the FY17 budget, can fund a new contract for four wings. People, tooling, non-recurring engineering and buildup of the 1st wing make up the preponderance of costs associated with this $123M, as detailed below:

Cost Element	Cost ($M)
1st article	$40.00
3 x LRIP ($12M/ea.)	$36.00
Tooling	$15.00
Startup	$29.50
Contract Total	$120.50
OGCs (PMO)	$3.40
Total	**$123.90**

173 A-10s—Fly existing A-10s wings to end of service life: Without additional funding, the Air Force could fly the A-10 fleet until the service limit of currently installed wings. Approximately 51 A-10s would be grounded by FY22, and flyable aircraft would decrease to 173 A-10s by the mid-2020s.

Acquire a new aircraft: The F-35A is designed to recapitalize the A-10 when it is retired.

The AX-2, a purpose built A-10 replacement, could also be considered among a range of options as A-10s reach their end of their service life, but is not funded.

The OA–X is a concept being evaluated through the Light Attack Experimentation Campaign and is not an A–10 replacement. OA–X could be an additive counter-land platform to complement the existing CAF. It is not a program at this time and would require additional TOA. [See page 19.]

QUESTIONS SUBMITTED BY MEMBERS POST HEARING

JUNE 7, 2017

QUESTIONS SUBMITTED BY MR. TURNER

Mr. TURNER. Approximately 6–7 weeks ago, you presented a briefing to a "HASC Roundtable" on the status of addressing the Physiological Episodes (PE) experienced by Navy and Marine Corps pilots flying F–18 and T–45 aircraft. Please provide us a brief update on the progress addressing this critical personnel safety issue.

Admiral GROSKLAGS. The root causes and solutions for PE prevention are being attacked in a coordinated effort between Naval Air System Command (NAVAIR), NASA, fleet operators, industry, international partners and aeromedical experts. No conclusive root cause or combination of causes yet has been identified.

T–45—The Navy's most recent efforts to mitigate physiological episodes center on alerting, protecting, preventing, and monitoring. Specific actions include performance of maintenance activities to ensure the hygiene and integrity of the breathing gas system and to functionally check and recertify critical systems sensors and components that affect its designed functionality; air quality is being measured on all aircraft. System modifications include addition of a water separator and a new oxygen pressure and concentration monitoring system. All flights include sorbent tube assemblies and hydrocarbon detectors worn by all aviators to measure the quality of the breathing gas reaching the aircrew mask. Additional PE sensors, data collection and analytics are being investigated and aggressively pursued, to include automated sensing monitoring and reporting technologies that measure aircraft performance and/or human performance in the flight environment. In parallel, NAVAIR continues to execute RCCA investigation through empowerment of a government, industry, and naval aviator team that is executing a disciplined process to eliminate or affirm potential causal factors on the basis of rigorous, data-driven, analytic efforts. Chief of Naval Air Training (CNATRA) Instructor Pilots (IPs) returned to flight using OBOGS on Monday 17 July. Student syllabus flights on OBOGS began the first week of August.

F/A–18—NAVAIR continues to support the F/A–18 RCCA investigation. The team consists of government, industry, and Royal Australian Air Force (RAAF) personnel working closely together to close out a 411 branch fault tree to determine the root cause of Physiological Events (PE). A toxicology team has been identified to understand the levels and potential impacts of contaminants in the F/A–18 Onboard Oxygen Generating System (OBOGS) and Environmental Control System (ECS), and the Navy is leveraging the experts in aeromedical and dive medicine to understand the effects of anomalous cockpit pressure situations on aircrew. All applicable and available aircraft, ECS and OBOGS data sources are being utilized to aid in PE analysis.

Mr. TURNER. What is DOD's search and rescue (SAR) capability in the U.S. Africa Command? Are there gaps in the capability or capacity for this theater that need to be addressed? Does the budget request or the unfunded requirements list address any requirements for SAR capability or capacity in U.S. Africa Command?

General DAVIS. [The information referred to was not available at the time of printing.]

Mr. TURNER. General Davis, what Marine aviation readiness concerns do you have? What are your concerns regarding spare parts for your tactical aircraft?

General DAVIS. [The information referred to was not available at the time of printing.]

Mr. TURNER. What is DOD's search and rescue (SAR) capability in the U.S. Africa Command? Are there gaps in the capability or capacity for this theater that need to be addressed? Does the budget request or the unfunded requirements list address any requirements for SAR capability or capacity in U.S. Africa Command?

Admiral MILLER. I defer to Joint Chiefs of Staff or Office of Secretary of Defense.

Mr. TURNER. You mentioned in your opening remarks that unmanned systems such as the MQ–4 Triton are modernization priorities for the fiscal year 2018 budget request. How does the budget request insure you meet the fielding of the capability for first deployment and Follow-on-Modernization.

Admiral MILLER. The fiscal year 2018 budget request supports the MQ–4C Triton fleet introduction as a key component to the Navy's "family of systems" to achieve maritime domain awareness. The program's fielding and modernization is aligned to support the Maritime Intelligence, Surveillance, Reconnaissance and Targeting (MISR&T) Transition Plan. This plan aligns key program events and funding with the development of the Multi-Intelligence (Multi-INT) capability upgrades and capacity to support warfighter demand. Triton will deliver capabilities essential for FY11 National Defense Authorization Act compliance. To support the MISR&T transition plan, the Navy prioritized remaining work and adjusted the Triton fielding plan by truncating the Triton Baseline program to deliver a safe, stable and effective system that establishes the foundation for Triton Multi-INT development. The PB18 request provides funding for an Early Operational Capability to facilitate Fleet introduction and learning in FY18 and will lay the foundation for Multi-INT fielding that supports the sundown of the EP-3. PB18 includes funding for retrofitting Low-Rate Initial Production (LRIP) lot 1 and 2 air vehicles from the Baseline configuration to the Multi-INT configuration. LRIP lot 1 and 2 vehicle retrofitting will increase Triton Multi-INT capacity from three to five air vehicles by the end of FY20.

Mr. TURNER. The subcommittee understands that the Air Force is undertaking a Lead Systems Integrator acquisition approach to re-host the capabilities of the C-130 Compass Call electronic warfare aircraft onto a new aircraft yet to be determined by the Lead Systems Integrator contractor. Does the Air Force plan to use a Lead Systems Integrator acquisition approach for the future recapitalization of other small inventory aircraft such as the E-3 AWACS or RC-135 Rivet Joint intelligence, surveillance, and reconnaissance platforms?

General BUNCH. The Air Force is employing a "System Integrator" approach for the COMPASS CALL re-host program. The term "Lead Systems Integrator" (LSI) is defined in the FY2008 NDAA, Sec. 802, as follows: "(A) a prime contractor for the development or production of a major system, if the prime contractor is not expected at the time of award to perform a substantial portion of the work on the system and the major subsystems; or (B) a prime contractor under a contract for the procurement of services the primary purpose of which is to perform acquisition functions closely associated with inherently governmental functions with respect to the development of a major system." The COMPASS CALL re-host program transfers existing mission equipment from the EC-130H aircraft to an existing commercial derivative aircraft platform, and the aircraft integration to be performed by L-3 Technologies makes up a significant portion of the COMPASS CALL re-host program. Moreover, L-3 is not being asked to perform any inherently governmental functions. Rather, the Air Force has defined the requirements of the COMPASS CALL re-host effort and the Air Force will ensure L-3's proposed solution meets those requirements. Thus, L-3 Technologies' role doesn't fit either prong of the LSI definition. The Air Force conducted extensive analysis in determining the most efficient, expedient and cost effective means to acquire COMPASS CALL capability through a system integrator approach. The Service will apply a similar level of rigor into development of the acquisition strategies for the follow-on capabilities replacing the E-3 AWACS and RC-135 Rivet Joint platforms. As for the requirements for the previously referenced platforms have yet to be determined, we cannot say with certainty what the acquisition strategies for these programs will be. Upon receipt of validated requirements, the Air Force's acquisition team will evaluate the efforts on a case by case basis and determine the most efficient, expedient and cost effective strategies for delivering these critical capabilities to the warfighter.

Mr. TURNER. What is DOD's search and rescue (SAR) capability in the U.S. Africa Command? Are there gaps in the capability or capacity for this theater that need to be addressed? Does the budget request or the unfunded requirements list address any requirements for SAR capability or capacity in U.S. Africa Command?

General HARRIS. DOD has several assets in theater providing SAR capabilities to U.S. Africa Command. The actual type, number and location are classified. There are shortfalls in requested forces for U.S. Africa Command. The Joint Staff is currently exploring courses of action to optimize SAR support across all Combatant Commands. Actual type, number and location of shortfalls for U.S. Africa Command are classified. There are contracted resources that supply SAR support for U.S. Africa Command that are included in DOD's budget request. Actual type, number and location of these contracted resources are classified. Beyond these contracted assets, no specific requests to support SAR capability or capacity in U.S. Africa Command are contained in the budget except as it relates to budget support for systems that support the overall SAR enterprise.

Mr. TURNER. Comparing the F-35A procurement in the 17 budget with the proposed 18 budget, it appears that the Air Force is requesting to increase the number of F-35s procured each year. Can you describe how many F-35s the Air Force is currently procuring each year and how many the department would like to buy?

General HARRIS. The FY 18 PB has the following procurement profile: FY18—46, FY19—48, FY20—48, FY 21—54, FY22—54. Additionally, the Air Force requested 14 on the Unfunded Priorities List and would like to procure 60 aircraft per year as quickly as possible.

Mr. TURNER. What funding is the Air Force requesting in the FY2018 President's Budget to support plans to extend the F–15C/D's service life? Additionally, the Air Force budget request cuts a significant amount of funding for installation of the Eagle Passive Active Warning Survivability System, or EPAWSS, on the F–15C aircraft. Why did this funding reduction occur and how does not installing EPAWSS on F–15C aircraft affect its future relevance and survivability in combat operations?

General HARRIS. The Fy18 PB requests $30.5M to begin a service life extension program for development/production of wings and longerons for the F–15. These actions maintain the necessary airframe economical service life options for the Air Force to finalize its force structure study. Current uncertainty regarding service life extension for the F–15C/D resulted in the reduction of F–15C/D EPAWSS procurement funding. The Air Force is fully funding the development of EPAWSS on both the F–15C/D and the F–15E. This is risk mitigation to maximize flexibility for future force structure decisions. Research and development of EPAWSS can be used across several platforms.

Mr. TURNER. How long is the Air Force planning to keep the E–8C JSTARS legacy fleet in service, and is the Air Force planning for a transition that does not result in a decrease of Moving Target Indicator intelligence capacity when the JSTARS Recapitalization aircraft is fielded?

General HARRIS. The Air Force will continue to assess E–8C service life, operational availability, and sustainment cots increases in conjunction with the JSTARS Recap fielding schedule to determine how and when to phase out the legacy fleet. Air Force senior leaders will brief potentional options to the Congressional Defense Committees as directed in the FY17 NDAA and Appropriations Act. Recapitalizing the E–8C fleet on a commercial derivative aircraft with an enhanced radar, modern battle management command and control suite, and robust communications is an available option to maintain the current Ground Moving Target Indicator capability. If a gap is unavoidable, the Air Force will consider all possible options to provide a similar capacity.

QUESTIONS SUBMITTED BY MS. TSONGAS

Ms. TSONGAS. Can you outline the additional capabilities that would be provided to the Navy by continued Block III advances to the F/A–18 Super Hornet and describe how these upgrades will benefit the Carrier Air Wing?

Admiral GROSKLAGS. As the primary Carrier Air Wing (CVW) weapons platform, the F/A–18 Super Hornet is complementary to the capabilities of the F–35 and E–2D and will optimize the capacity of the CVW of the future. The Block III configuration specifically increases F/A–18 situational awareness, aircraft survivability and extends its range. Specific advancements increase battlespace awareness, range, survivability and lethality:

—Active Electronically Scanned Array (AESA) radar upgrades, Satellite Communications (SATCOM) and Infrared Search and Track (IRST)—improves lethality.
—New aircrew display—battlespace awareness
—New Digital Targeting Processing Network (DTP–N) controller and Tactical Targeting Network Technology (TTNT) card—advances computing, high speed network transfer and battlespace awareness.
—Conformal Fuel Tanks (CFTs)—increases range while reducing aircraft signature and increases lethality by maximizing Super Hornet weapons capacity.
—Integrated Defensive Electronic Counter Measures (IDECM) Block IV Suite—increases aircraft survivability.

F/A–18E/F Block III will be delivered with a 9K hour frame—keeps these assets on the flight line and precludes the need for a costly Service Life Extension Program (SLEP).

Ms. TSONGAS. Can you outline the additional capabilities that would be provided to the Navy by continued Block III advances to the F/A–18 Super Hornet and describe how these upgrades will benefit the Carrier Air Wing?

General HARRIS. The Air Force refers to the Navy.

QUESTION SUBMITTED BY MR. GAETZ

Mr. GAETZ. Does the U.S. Air Force place strategic importance on the Block Four capability of the F–35?

General BUNCH and General HARRIS. Yes, Block 4 is very important. The Air Force cannot emphasize enough how important it is that we fully fund Block 4 to prevent delaying required capabilities for American and Coalition warfighters, including integration of additional weapons and upgrades to mission systems as the electronic warfare system, data link systems, and radar that will ensure operational advantage against the emerging 2025 threat.

QUESTIONS SUBMITTED BY MR. BACON

Mr. BACON. The information needs of the joint force will require that every platform is a sensor, especially those designed to operate in contested or denied areas. The F–35 will be deployed in large numbers and will have one of the most capable suites of sensors ever put on an airplane. What current efforts has your service undertaken to ensure the joint force can access and integrate the data collected by F–35 sensors?

Admiral GROSKLAGS. The Navy and Marine Corps Team is working on solutions to ensure in-flight sensor data is transmitted to the carrier intelligence centers via appropriate radio datalinks. Post-flight sensor data will be recorded for download, then transmitted throughout Navy ship networks and Distributed Common Ground System (DCGS) for processing, exploitation, and dissemination analysis to the Joint Force. Because F–35 post-flight mission products are classified at the SAP level, the Team is pursuing the implementation of a cross-domain solution to move the appropriate information across classification boundaries. The Link-16 network will be the foundation of the tactical airborne data links for decades, and the primary means by which the F–35 will be integrated with other naval aviation assets (E–2, F/A–18, EA–18, etc) in the NIFC–CA kill chain. Link-16's value is breadth of deployment across the multi-national force, but it is not adequate for sharing all data in all scenarios. There is still a need for specialized waveforms and networks which may require a gateway solution for transfer and fusion of data (MADL, TTNT, and CEC).

Mr. BACON. The information needs of the joint force will require that every platform is a sensor, especially those designed to operate in contested or denied areas. The F–35 will be deployed in large numbers and will have one of the most capable suites of sensors ever put on an airplane. What current efforts has your service undertaken to ensure the joint force can access and integrate the data collected by F–35 sensors?

Admiral MILLER. The F-35's sensor fusion solution and data sharing capabilities are focused on providing the interoperability required by the warfighter in support of the execution of the mission at the tactical level. The program is currently planning increased capability in these areas as part of Follow-on Modernization, to include Tactical Data Recording capability, which will allow the warfighter to record and use this data for "next day" missions. While there is no current capability or approved operational requirement to contribute to the Process, Exploit, Dissemination (PED) architecture, the Services continue to investigate future opportunities to include this capability in future F-35 upgrades.

Mr. BACON. Given the current threat environment, can you explain the USAF's intent to drastically reduce development and procurement funding for the F–15 C/D EPAWSS self-protection upgrade?

General BUNCH and General HARRIS. The F–15E has more service life remaining that the F–15C/D, and the Air Force plans F–15E sustainment into the 2040s. The F–15E also operates in air-to-surface environments where electronic warfare self-protection is an absolute must. The Air Force is funding procurement of Eagle Passive Active Warning Survivability System (EPAWSS) for the F–15E based on the projected military utility and return on investment. Current uncertainty regarding service life extension for the F–15C/D resulted in the reduction of F–15C/D EPAWSS procurement funding.

The Air Force is fully funding the development of EPAWSS on both the F–15C/D and the F–15E. This is risk mitigation to maximize flexibility for future structure decisions. The FY18 PB fully funds development (R&D) of EPAWSS Increment 1, which can be used across several platforms.

Mr. BACON. The E–8 JSTARS remains in critical demand by Combatant Commanders around the world. What is the manpower and equipment cost required to establish standing E–8 forward operating locations in Europe and the Pacific, like we have done successful for decades with the E–3 AWACS and the RC–135 programs?

General BUNCH and General HARRIS. The Air Force has not conducted a full cost analysis for these two locations. However, initial analysis indicates a rough estimate for costs to establish two JSTARS FOLs would include initial upfront costs of approximately $200M (MILCON and mx equipment) and annual combined sustainment costs of approximately $50M (O&M, manpower, and annual TDY/Transportation costs at both locations. These values are very rough but demonstrate the significant upfront costs required to execute. Costs aside, Air Force would find it extremely difficult to execute from an operations perspective because of the low Aircraft Availability rates driven by increased Primary Depot Maintenance backlogs.

The AF understands the value of a continuing presence of JSTARS in multiple theaters and continues to use the GFMAP to respond to Combatant Commanders' highest priority requirements.

Mr. BACON. The information needs of the joint force will require that every platform is a sensor, especially those designed to operate in contested or denied areas. The F–35 will be deployed in large numbers and will have one of the most capable suites of sensors ever put on an airplane. What current efforts has your service undertaken to ensure the joint force can access and integrate the data collected by F–35 sensors?

General BUNCH and General HARRIS. The F–35's sensor fusion solution and data sharing capabilities are focused on providing the interoperability required by the warfighter in support of the execution of the mission at the tactical level. The program is currently planning increased capability in these areas as part of Follow-on Modernization, to include Tactical Data Recording capability, which will allow the warfighter to record and use this data for "next day" missions. While there is no current capability or approved operational requirement to contribute to the Process, Exploit, Dissemination (PED) architecture, the Services continue to investigate future opportunities to include this capability in future F–35 upgrades.

www.ingramcontent.com/pod-product-compliance
Lightning Source LLC
Chambersburg PA
CBHW062217220526
45471CB00009B/3243